Multi-Carrier
Digital Communications
Theory and Applications of OFDM

Information Technology: Transmission, Processing, and Storage

Multi-Carrier Digital Communications: Theory and Applications of OFDM
Ahmad R. S. Bahai and Burton R. Saltzberg

Principles of Digital Transmission: With Wireless Applications
Sergio Benedetto and Ezio Biglieri

Simulation of Communication Systems, 2nd Edition: Methodology, Modeling, and Techniques
Michel C. Jeruchim, Philip Balaban, and K. Sam Shanmugan

A Continuation Order Plan is available for this series. A continuation order will bring delivery of each new volume immediately upon publication. Volumes are billed only upon actual shipment. For further information please contact the publisher.

Multi-Carrier
Digital Communications
Theory and Applications of OFDM

Ahmad R. S. Bahai and
Burton R. Saltzberg

Algorex, Inc.
Iselin, New Jersey

Kluwer Academic / Plenum Publishers
New York, Boston, Dordrecht, London, Moscow

ISBN: 0-306-46296-6

©1999 Kluwer Academic / Plenum Publishers, New York
233 Spring Street, New York, N.Y. 10013

10 9 8 7 6 5 4 3 2 1

A C.I.P. record for this book is available from the Library of Congress

Printed in the United States of America

Preface

Multi-carrier modulation, in particular Orthogonal Frequency Division Multiplexing (OFDM), has been successfully applied to a wide variety of digital communications applications over the past several years. Although OFDM has been chosen as the physical layer standard for a diversity of important systems, the theory, algorithms, and implementation techniques remain subjects of current interest. This is clear from the high volume of papers appearing in technical journals and conferences.

This book is intended to be a concise summary of the present state of the art of the theory and practice of OFDM technology. The authors believe that the time is ripe for such a treatment. Particularly based on one of the author's long experience in development of wireless systems, and the other's in wireline systems, we have attempted to present a unified presentation of OFDM performance and implementation over a wide variety of channels. It is hoped that this will prove valuable both to developers of such systems and to researchers and graduate students involved in analysis of digital communications.

In the interest of brevity, we have minimized treatment of more general communication issues. There exist many excellent texts on communication theory and technology. Only brief summaries of topics not specific to multi-carrier modulation are presented in this book where essential.

We begin with a historical overview of multi-carrier communications, wherein its advantages for transmission over highly dispersive channels have long been recognized, particularly before the development of equalization techniques. We then focus on the bandwidth efficient technology of OFDM, in particular the digital signal processing techniques that have made the modulation format practical. Several chapters describe and analyze the sub-systems of an OFDM implementation, such as synchronization, equalization, and coding. Analysis of performance over channels with various impairments is presented. The chapter on effects of clipping presents results of the authors that have not yet been published elsewhere.

The book concludes with descriptions of three very important and diverse applications of OFDM that have been standardized and are now being deployed. ADSL provides access to digital services at several Mb/s over the ordinary wire-pair connection between customers and the local telephone company central office. Digital Broadcasting enables the radio reception of high quality digitized sound and video. A unique configuration that is enabled by OFDM is the simultaneous transmission of identical signals by geographically dispersed transmitters. Finally, the new development of wireless LANs for multi-Mb/s communications is presented. Each of these successful applications required the development of new fundamental technology.

Multi-carrier modulation continues to evolve rapidly. It is hoped that this book will remain a valuable summary of the technology, providing an understanding of new advances as well as the present core technology.

We acknowledge the extensive review and many valuable suggestions of Professor Kenji Kohiyama, our former colleagues at AT&T Bell Laboratories and colleagues at Algorex. Gail Bryson performed the very difficult task of editing and assembling this text. The continuing support of

Kambiz Homayounfar was essential to its completion. Last, but by no means least, we are thankful to our families for their support and patience.

Contents

CHAPTER 1 INTRODUCTION TO DIGITAL COMMUNICATIONS 2

1.1   BACKGROUND 2
1.2   EVOLUTION OF OFDM 7

CHAPTER 2 SYSTEM ARCHITECTURE..17

2.1   MULTI-CARRIER SYSTEM FUNDAMENTALS 17
2.2   DFT 20
2.3   PARTIAL FFT 25
2.4   CYCLIC EXTENSION 27
2.5   CHANNEL ESTIMATION...29
2.6   APPENDIX — MATHEMATICAL MODELLING OF OFDM FOR TIME-VARYING
RANDOM CHANNEL..32

CHAPTER 3 PERFORMANCE OVER TIME-INVARIANT CHANNELS 41

3.1   TIME-INVARIANT NON-FLAT CHANNEL WITH COLORED NOISE...............41
3.2   ERROR PROBABILITY..42
3.3   BIT ALLOCATION..46
3.4   BIT AND POWER ALLOCATION ALGORITHMS FOR FIXED BIT RATE 53

CHAPTER 4 CLIPPING IN MULTI-CARRIER SYSTEMS 57

4.1   INTRODUCTION 57
4.2   POWER AMPLIFIER NON-LINEARITY 59
4.3   BER ANALYSIS..63
4.4   BANDWIDTH REGROWTH..76

CHAPTER 5 SYNCHRONIZATION 83

5.1 TIMING AND FREQUENCY OFFSET IN OFDM 83
5.2 SYNCHRONIZATION AND SYSTEM ARCHITECTURE 88
5.3 TIMING AND FRAME SYNCHRONIZATION...89
5.4 FREQUENCY OFFSET ESTIMATION 91
5.5 PHASE NOISE 93

CHAPTER 6 EQUALIZATION 103

6.1 INTRODUCTION 103
6.2 TIME DOMAIN EQUALIZATION 104
6.3 EQUALIZATION IN DMT 109
6.4 FREQUENCY DOMAIN EQUALIZATION 116
6.5 ECHO CANCELLATION 120
6.6 APPENDIX — JOINT INNOVATION REPRESENTATION OF ARMA MODELS............127

CHAPTER 7 CHANNEL CODING 135

7.1 NEED FOR CODING..135
7.2 BLOCK CODING IN OFDM 136
7.3 CONVOLUTIONAL ENCODING 142
7.4 CONCATENATED CODING 147
7.5 TRELLIS CODING IN OFDM 148
7.6 TURBO CODING IN OFDM...153

CHAPTER 8 ADSL...159

8.1 WIRED ACCESS TO HIGH RATE DIGITAL SERVICES..159
8.2 PROPERTIES OF THE WIRE-PAIR CHANNEL 160
8.3 ADSL SYSTEMS...170

CHAPTER 9 WIRELESS LAN 175

9.1 INTRODUCTION 175
9.2 PHYSICAL LAYER TECHNIQUES FOR WIRELESS LAN....................................181
9.3 OFDM FOR WIRELESS LAN...182
9.4 RECEIVER STRUCTURE 187

CHAPTER 10 DIGITAL BROADCASTING..191

10.1   BROADCASTING OF DIGITAL AUDIO SIGNALS 191
10.2   SIGNAL FORMAT...194
10.3   OTHER DIGITAL BROADCASTING SYSTEMS 197
10.4   DIGITAL VIDEO BROADCASTING 198

CHAPTER 11 FUTURE TRENDS..203

11.1   COMPARISON WITH SINGLE CARRIER MODULATION 203
11.2   MITIGATION OF CLIPPING EFFECTS...205
11.3   OVERLAPPED TRANSFORMS 206
11.4   COMBINED CDMA AND OFDM 210
11.5   ADVANCES IN IMPLEMENTATION..213

INDEX...217

Multi-Carrier
Digital Communications
Theory and Applications of OFDM

Chapter 1 *Introduction to Digital Communications*

1.1 Background

The physical layer of digital communications includes mapping of digital input information into a waveform for transmission over a communication channel, which may introduce various forms of distortion as well as noise, and mapping the received waveform into digital information that hopefully agrees with the original input [1]. The simplest form of such communication, as least conceptually, is Pulse Amplitude Modulation (PAM), shown in Figure 1.1. Here the transmitted waveform is of the form

$$s(t) = \sum_n a_n \, g(t - nT) \qquad\qquad 1.1$$

where the information to be transmitted is given by the sequence of a_ns, $1/T$ is the symbol rate, and $g(t)$ is the impulse response of the transmit filter, usually low-pass. The a_ns are chosen from an alphabet of size L, so the bit rate is $1/T \log_2 L$. It is desirable that the alphabet be both zero mean and equally spaced. The values of a_n can be written as

$$\{-A(L-1),\cdots,-A, A,\cdots, A(L-1)\}. \qquad 1.2$$

Assuming the a_n s are equiprobable, the transmitted power is

$$A^2 \frac{L^2-1}{3T} \int_{-\infty}^{\infty} g^2(t)\, dt. \qquad 1.3$$

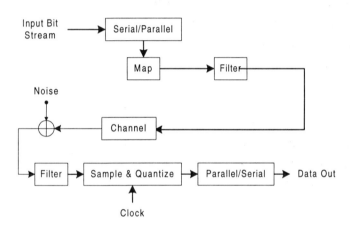

Figure 1.1. A basic PAM system.

At the receiver, the signal is filtered by $r(t)$, which may include an adaptive equalizer (sampled), and the nearest permitted member of the alphabet is output. In order to avoid inter-symbol interference, it is desirable that $x(t) = g(t) * h(t) * r(t) = 0$ for all $t = kT$, k an integer $\neq 0$, where $h(t)$ is the channel impulse response. This is the Nyquist criterion, which is given in the frequency domain by:

$$\sum_m X(f + \frac{m}{T}) = const. \qquad 1.4$$

The minimum bandwidth required is $1/2T$. This is met by a frequency response that is constant for $-1/2T < f < 1/2T$, whose corresponding time response is

$$x(t) = \frac{\sin \pi t / T}{\pi t / T}. \qquad 1.5$$

Some excess bandwidth, denoted by the roll-off factor, is desirable in order for the time response to decay more quickly. Note that $r(t)$ is not a matched filter, because it must satisfy the inter-symbol interference constraint.

If $r(t)$ has gain such that the alphabet levels of $x(0)$ are also spaced by $2A$, then errors will occur when the noise at the sampler satisfies $|n| > A$ for interior levels, or $n > A$ or $n < -A$ for the outer levels. If the noise is Gaussian with power spectral density $N(f)$ at the receiver input, then the noise variance is:

$$\sigma^2 = \int_{-\infty}^{\infty} N(f) |R(f)|^2 \, df \qquad 1.6$$

and the error probability per symbol is

$$P_e = \frac{2(L-1)}{L} Q(\frac{A}{\sigma}), \qquad 1.7$$

where

$$Q(x) = \frac{1}{\sqrt{2\pi}} \int_x^{\infty} e^{-y^2/2} \, dy \qquad 1.8$$

is the normal error integral.

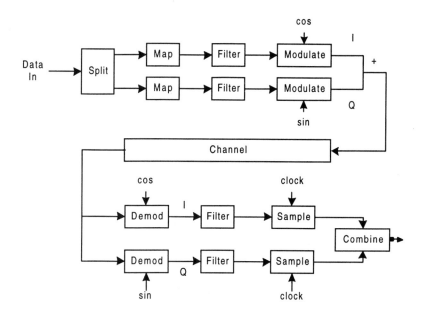

Figure 1.2. A basic QAM system.

PAM is only suitable over channels that exist down to, but might not necessarily include, zero frequency. If zero frequency is absent, a modulation scheme that puts the signal spectrum in the desired frequency band is required. Of particular interest, both in its own right and as a component of OFDM, is Quadrature Amplitude Modulation (QAM). The simplest form of QAM, shown in Figure 1.2, may be thought of as two PAM signals, modulated by carriers at the same frequency but 90 degrees out of phase. At the receiver, demodulation by the same carriers separates the signal components. Unlike some other modulation schemes, such as FM, QAM is bandwidth efficient in that it requires the same bandwidth as a PAM signal of the same bit rate. Furthermore, the performance of QAM in noise is comparable to that of PAM.

The QAM line signal is of the form

$$\sum_n a_n \, g(t-nT)\cos\omega t - \sum_n b_n \, g(t-nT)\sin\omega t. \qquad\qquad 1.9$$

This line signal may also be written in the form of:

$$\mathrm{Re}\left\{\sum_n c_n \, g(t-nT)\,e^{j\omega t}\right\}, \qquad\qquad 1.10$$

where the pair of real symbols a_n and b_n are treated as a complex symbol $c_n = a_n + jb_n$. The required bandwidth for transmitting such complex symbols is $1/T$. The complex symbol values are shown as a "constellation" in the complex plane. Figure 1.3 shows the constellation of a 16-point QAM signal, which is formed from 4-point PAM.

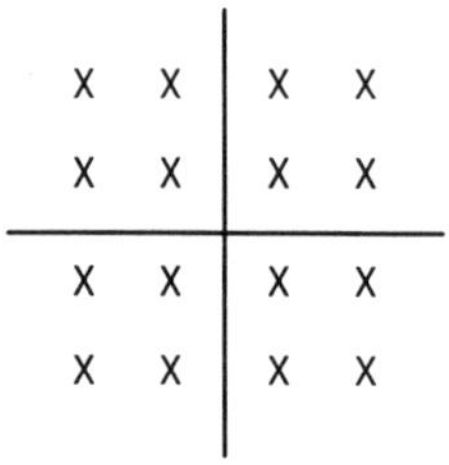

Figure 1.3. A QAM constellation.

It is not necessary that the constellation be square. Figure 1.4 shows how input information can be mapped arbitrarily into constellation points. A constellation with a more circular boundary provides better noise performance. By grouping n successive complex symbols as a unit, we can treat such units as symbols in $2n$-dimensional space. In this case, Figure 1.4 can be extended to include a large enough serial-to-parallel converter that accommodates the total number of bits in n symbols, and a look-up table with $2n$ outputs.

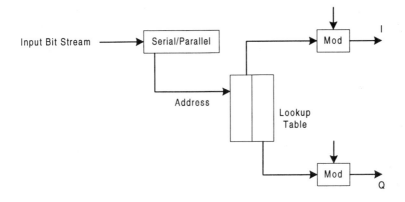

Figure 1.4. General form of QAM generation.

1.2 Evolution of OFDM

The use of Frequency Division Multiplexing (FDM) goes back over a century, where more than one low rate signal, such as telegraph, was carried over a relatively wide bandwidth channel using a separate carrier frequency for each signal. To facilitate separation of the signals at the receiver, the carrier frequencies were spaced sufficiently far apart so that the signal spectra did not overlap. Empty spectral regions between the signals assured that they could be separated with readily realizable filters. The resulting spectral efficiency was therefore quite low.

Instead of carrying separate messages, the different frequency carriers can carry different bits of a single higher rate message. The source may be in such a parallel format, or a serial source can be presented to a serial-to-parallel converter whose output is fed to the multiple carriers.

Such a parallel transmission scheme can be compared with a single higher rate serial scheme using the same channel. The parallel system, if built straightforwardly as several transmitters and receivers, will certainly be more costly to implement. Each of the parallel sub-channels can carry a low signalling rate, proportional to its bandwidth. The sum of these signalling rates is less than can be carried by a single serial channel of that combined bandwidth because of the unused guard space between the parallel sub-carriers. On the other hand, the single channel will be far more susceptible to inter-symbol interference. This is because of the short duration of its signal elements and the higher distortion produced by its wider frequency band, as compared with the long duration signal elements and narrow bandwidth in sub-channels in the parallel system. Before the development of equalization, the parallel technique was the preferred means of achieving high rates over a dispersive channel, in spite of its high cost and relative bandwidth inefficiency. An added benefit of the parallel technique is reduced susceptibility to most forms of impulse noise.

The first solution of the bandwidth efficiency problem of multi-tone transmission (not the complexity problem) was probably the "Kineplex" system. The Kineplex system was developed by Collins Radio Co. [2] for data transmission over an H.F. radio channel subject to severe multi-path fading. In that system, each of 20 tones is modulated by differential 4-PSK without filtering. The spectra are therefore of the $sin(kf)/f$ shape and strongly overlap. However, similar to modern OFDM, the tones are spaced at frequency intervals almost equal to the signalling rate and are capable of separation at the receiver.

The reception technique is shown in Figure 1.5. Each tone is detected by a pair of tuned circuits. Alternate symbols are gated to one of the tuned circuits, whose signal is held for the duration of the next symbol. The signals in the two tuned circuits are then processed to determine their phase difference, and therefore the transmitted information. The older of the two signals is then quenched to allow input of the next symbol. The key to the success of the technique is that the time response of each tuned circuit to all tones, other than the one to which it is tuned, goes through zero at the end of the gating interval, at which point that interval is equal to the reciprocal of the frequency separation between tones. The gating time is made somewhat shorter than the symbol period to reduce inter-symbol interference, but efficiency of 70% of the Nyquist rate is achieved. High performance over actual long H.F. channels was obtained, although at a high implementation cost. Although fully transistorized, the system required two large bays of equipment.

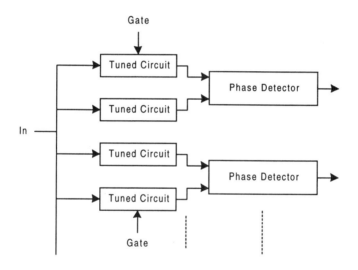

Figure 1.5. The Collins Kineplex receiver.

A subsequent multi-tone system [3] was proposed using 9-point QAM constellations on each carrier, with correlation detection employed in the receiver. Carrier spacing equal to the symbol rate provides optimum spectral efficiency. Simple coding in the frequency domain is another feature of this scheme.

The above techniques do provide the orthogonality needed to separate multi-tone signals spaced by the symbol rate. However the *sin(kf)/f* spectrum of each component has some undesirable properties. Mutual overlap of a large number of sub-channel spectra is pronounced. Also, spectrum for the entire system must allow space above and below the extreme tone frequencies to accommodate the slow decay of the sub-channel spectra. For these reasons, it is desirable for each of the signal components to be bandlimited so as to overlap only the immediately adjacent sub-carriers, while remaining orthogonal to them. Criteria for meeting this objective are given in References [4] and [5].

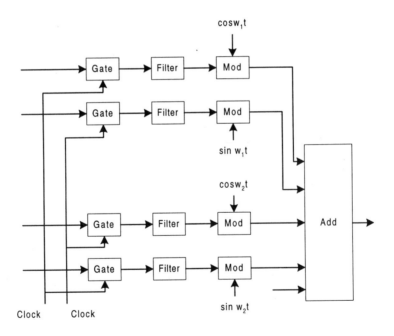

Figure 1.6. An early version of OFDM.

In Reference [6] it was shown how bandlimited QAM can be employed in a multi-tone system with orthogonality and minimum carrier spacing (illustrated in Figure 1.6). Unlike the non-bandlimited OFDM, each carrier must carry Staggered (or Offset) QAM, that is, the input to the I and Q modulators must be offset by half a symbol period. Furthermore, adjacent carriers must be offset oppositely. It is interesting to note that Staggered QAM is identical to Vestigial Sideband (VSB) modulation. The low-pass filters $g(t)$ are such that $G^2(f)$, the combination of transmit and receive filters, is Nyquist, with the roll-off factor assumed to be less than 1.

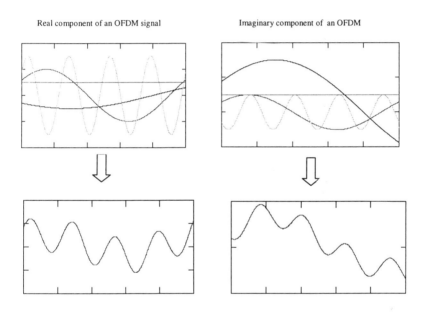

Real component of an OFDM signal   Imaginary component of an OFDM

Figure 1.7. OFDM modulation concept: Real and Imaginery components of an OFDM symbol is the superposition of several harmonics modulated by data symbols.

The major contribution to the OFDM complexity problem was the application of the Fast Fourier Transform (FFT) to the modulation and demodulation processes [7]. Fortunately, this occurred at the same time digital signal processing techniques were being introduced into the design of modems. The technique involved assembling the input information into blocks of *N* complex numbers, one for each sub-channel. An inverse FFT is performed on each block, and the resultant transmitted serially.At the receiver, the information is recovered by performing an FFT on the received block of signal samples. This form of OFDM is often referred to as Discrete Multi-Tone (DMT). The spectrum of the signal on the line is identical to that of *N* separate QAM signals, at *N* frequencies separated by the signalling rate. Each such QAM signal carries one of the original input complex numbers. The spectrum of each QAM signal is of the form $\sin(kf)/f$, with nulls at the center of the other sub-carriers, as in the earlier OFDM systems, and as shown in Figure 1.8 and Figure 1.9.

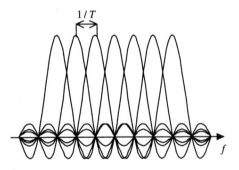

Figure 1.8. Spectrum overlap in OFDM.

A block diagram of a very basic DMT system is shown Figure 1.10. Several critical blocks are not shown. As described more thoroughly in Chapter 2, care must be taken to avoid overlap of consecutive transmitted blocks, a problem that is solved by the use of a cyclic prefix. Another issue is how to transmit the sequence of complex numbers from the output of the inverse FFT over the channel.

The process is straightforward if the signal is to be further modulated by a modulator with I and Q inputs.

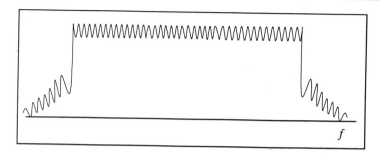

Figure 1.9. Spectrum of OFDM signal.

Otherwise, it is necessary to transmit real quantities. This can be accomplished by first appending the complex conjugate to the original input block. A $2N$-point inverse FFT now yields $2N$ real numbers to be transmitted per block, which is equivalent to N complex numbers.

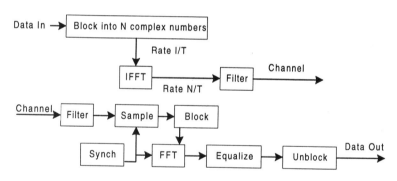

Figure 1.10. Very basic OFDM system.

The most significant advantage of this DMT approach is the efficiency of the FFT algorithm. An N-point FFT requires only on the order of $N \log N$ multiplications, rather than N^2 as in a straightforward computation. The efficiency is particularly good when N is a power of 2, although that is not generally necessary. Because of the use of the FFT, a DMT system typically requires fewer computations per unit time than an equivalent single channel system with equalization. An overall cost comparison between the two systems is not as clear, but the costs should be approximately equal in most cases. It should be noted that the bandlimited system of Figure 1.6 can also be implemented with FFT techniques [8], although the complexity and delay will be greater than DMT.

Over the last 20 years or so, OFDM techniques and, in particular, the DMT implementation, has been used in a wide variety of applications [9]. Several OFDM voiceband modems have been introduced, but did not succeed commercially because they were not adopted by standards bodies. DMT has been adopted as the standard for the Asymmetric Digital Subscriber Line (ADSL), which provides digital communication at several Mb/s from a telephone company central office to a subscriber, and a lower rate in the reverse direction, over a normal twisted pair of wires in the loop plant.

OFDM has been particularly successful in numerous wireless applications, where its superior performance in multi-path environments is desirable. Wireless receivers detect signals distorted by time and frequency selective fading. OFDM in conjunction with proper coding and interleaving is a powerful technique for combating the wireless channel impairments that a typical OFDM wireless system might face, as is shown in Figure 1.11.

A particularly interesting configuration, discussed in Chapter 10, is the Single Frequency Network (SFN) used for broadcasting of digital audio or video signals. Here many geographically separated transmitters broadcast identical and synchronized signals to cover a large region. The reception of such signals by a receiver is equivalent to an extreme form of multi-path. OFDM is the technology that makes this configuration viable.

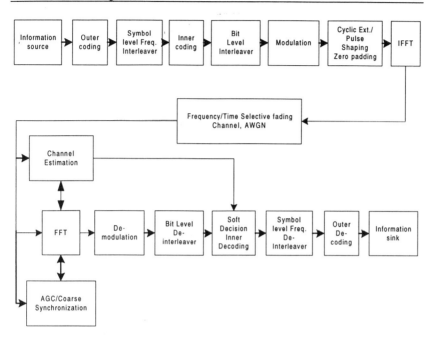

Figure 1.11. A Typical Wireless OFDM architecture.

Another wireless application of OFDM is in high speed local area networks (LANs). Although the absolute delay spread in this environment is low, if very high data rates, in the order of many tens of Mb/s, is desired, then the delay spread may be large compared to a symbol interval. OFDM is preferable to the use of long equalizers in this application.

It is expected that OFDM will be applied to many more new communications systems over the next several years.

References

1. Gitlin, R.D., Hayes J.F., Weinstein S.B. *Data Communications Principles.* New York: Plenum, 1992.

2. Doelz, M.L., Heald E.T., Martin D.L. "Binary Data Transmission Techniques for Linear Systems." *Proc. I.R.E.*; May 1957; 45: 656-661.

3. Franco, G.A., Lachs G. "An Orthogonal Coding Technique for Communications." *I. R. E. Int. Conv. Rec.;* 1961; 8: 126-133.

4. Chang, R.W. "Synthesis of Band-Limited Orthogonal Signals for Multichannel Data Transmission." *Bell Sys. Tech. J.*; Dec 1966; 45: 1775-1796.

5. Shnidman, D.A. "A Generalized Nyquist Criterion and an Optimum Linear Receiver for a Pulse Modulation System." *Bell Sys. Tech. J.*; Nov 1966; 45: 2163-2177.

6. Saltzberg, B.R. "Performance of an Efficient Parallel Data Transmission System." *IEEE Trans. Commun.*; Dec 1967; COM-15; 6: 805-811.

7. Weinstein, S.B., Ebert P.M. "Data Transmission By Frequency Division Multiplexing Using the Discrete Fourier Transform." *IEEE Trans. Commun.*, Oct 1971; COM-19; 5: 628-634.

8. Hirosaki, B. "An Orthogonally Multiplexed QAM System Using the Discrete Fourier Transform." *IEEE Trans. Commun.*; Jul 1981; COM-29; 7: 982-989.

9. Bingham, J.A.C. "Multicarrier Modulation for Data Transmission: An Idea Whose Time Has Come." *IEEE Commun. Mag.,* May 1990; 28:   14.

Chapter 2 *System Architecture*

This chapter presents a general overview of system design for multi-carrier modulation. First, a review of the OFDM system is discussed, then major system blocks will be analyzed.

2.1 Multi-Carrier System Fundamentals

Let $[D_0, D_1, \cdots, D_{N-1}]$ denote data symbols. Digital signal processing techniques, rather than frequency synthesizers, can be deployed to generate orthogonal sub-carriers. The DFT as a linear transformation maps the complex data symbols $[D_0, D_1, \cdots, D_{N-1}]$ to OFDM symbols $[d_0, d_1, \cdots, d_{N-1}]$ such that

$$d_k = \sum_{n=0}^{N-1} D_n e^{j2\pi n \frac{k}{N}}. \qquad 2.1$$

The linear mapping can be represented in matrix form as:

$$\overline{d} = \overline{W}\,\overline{D}, \qquad 2.2$$

17

where:

$$\overline{W} = \begin{bmatrix} 1 & \cdots & 1 & 1 \\ 1 & W & \cdots & W^{N-1} \\ 1 & W^2 & & W^{2(N-1)} \\ & \cdots & & \cdots \\ 1 & W^{N-1} & & W^{N(N-1)} \end{bmatrix}.$$

                                                                                                                    2.3

and,

$$W = e^{j2\pi/N}.$$ 2.4

$\overline{W}$ is a symmetric and orthogonal matrix. After FFT, a cyclic pre/postfix of lengths k_1 and k_2 will be added to each block (OFDM symbol) followed by a pulse shaping block. Proper pulse shaping has an important effect in improving the performance of OFDM systems in the presence of some channel impairments, and will be discussed in Chapter 5. The output of this block is fed to a D/A at the rate of f_s, and low-pass filtered. A basic representation of the equivalent complex baseband transmitted signal is

$$x(t) = \sum_{n=0}^{N-1} \left\{ D_n e^{j2\pi \frac{n}{N} f_s t} \right\}$$ 2.5

for

$$-\frac{k_1}{f_s} < t < \frac{N+k_2}{f_s}.$$ 2.6

A more accurate representation of OFDM signal including windowing effect is

$$x(t) = \sum_{l=-\infty}^{\infty} \sum_{k=-k_1}^{N+k_2} \sum_{n=0}^{N-1} \left\{ D_{nl} e^{j2\pi \frac{n}{N} k} \right\} w\left(t - \frac{k}{f_s} - lT\right)$$ 2.7

D_{nl} represents the nth data symbol transmitted during the l-th OFDM block, $T = (N + k_1 + k_2)/f_s$ is the OFDM block duration, and $w(t)$ is the window or pulse shaping function. The extension of the OFDM block is equivalent to adding a cyclic pre/postfix in the discrete domain. The received signal for a time-varying random channel is

$$r(t) = \int_0^\infty x(t-\tau)h(t,\tau)d\tau + n(t).$$ 2.8

The received signal is sampled at $t = k / f_s$ for $k = \{-k_1, \ldots N + k_2 - 1\}$. With no inter-block interference, and assuming[1] that the windowing function satisfies $w(n-l) = \delta_{nl}$, the output of the FFT block at the receiver is

$$\tilde{D}_m = \frac{1}{N} \sum_{k=0}^{N-1} r_k e^{-j2\pi m \frac{k}{2N}},$$ 2.9

where

$$r_k = \sum_{n=0}^{N-1} H_n D_n e^{j2\pi \frac{n}{2N}k} + n(k).$$ 2.10

A complex number H_n is the frequency response of the time-invariant channel $h(t-\tau)$ at frequency n / T. So,

$$\tilde{D}_m = \begin{cases} H_n D_n + N(n), & n = m \\ N(n), & n \neq m \end{cases}$$ 2.11

$n(t)$ is white Gaussian noise with a diagonal covariance matrix of $E(n(k)n(l)) = \delta I$. Therefore, the noise components for different sub-carriers are not correlated,

$$E(\bar{n}(k)\bar{n}^*(l)) = W\delta I W^T = \delta I.$$ 2.12

where $\bar{n}(k)$ is the vector of noise samples $n(k), \ldots, n(k\text{-}N)$.

A detailed mathematical analysis of OFDM in multi-path Rayleigh fading is presented in the Appendix.

2.2 DFT

The key components of an OFDM system are the Inverse DFT in the transmitter and the DFT in the receiver. These components must implement

$$d_n = \frac{1}{\sqrt{N}} \sum_{k=0}^{N-1} D_k W_N^{kn}, \ n = 0,...,N-1 \qquad 2.13$$

and,

$$D_n = \frac{1}{\sqrt{N}} \sum_{k=0}^{N-1} d_k W_N^{-kn}, \ n = 0,...,N-1 \qquad 2.14$$

respectively.

These operations perform reversible linear mappings between N complex data symbols $\{D\}$ and N complex line symbols $\{d\}$. It can be seen that the two operations are essentially identical.

The scale factor $1/\sqrt{N}$ provides symmetry between the operations and also preservation of power. Frequently, a scale factor of $1/N$ is used in one direction and unity in the other instead. In an actual implementation, this is immaterial because scaling is chosen to satisfy considerations of overflow and underflow rather than any mathematical definition.

A general N-to-N point linear transformation requires N^2 multiplications and additions. This would be true of the DFT and IDFT if each output symbol were calculated separately. However, by calculating the outputs simultaneously and taking advantage of the cyclic properties of the multipliers $e^{\pm j2\pi nk/N}$, Fast Fourier Transform (FFT) techniques reduce the number of computations to the order of $N \log N$. The FFT is most efficient when N is a power of two. Several variations of the FFT exist, with different ordering of the inputs and outputs, and different use of temporary memory. One variation, decimation in time, is shown below.

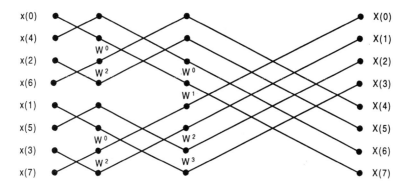

Figure 2.1. An FFT implementation (decimation in time).

Figure 2.2 shows the architecture of an OFDM system capable of using a further stage of modulation employing both in-phase and quadrature modulators. This configuration is common in wireless communication systems for modulating baseband signals to the required IF or RF frequency band. It should be noted that the basic configuration illustrated does not account for channel dispersion, which is almost always present. The channel dispersion problem is solved by using the cyclic prefix which will be described later.

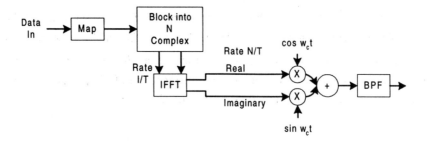

TRANSMITTER

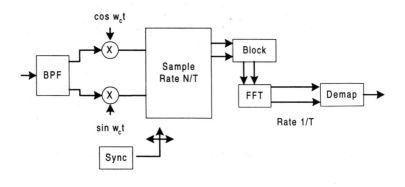

RECEIVER

Figure 2.2. System with complex transmission.

Small sets of input bits are first assembled and mapped into complex numbers which determine the constellation points of each sub-carrier. In most wireless systems, smaller constellation is formed for each sub-carrier. In wireline systems, where the signal-to-noise ratio is higher and variable across the frequency range, the number of bits assigned to each sub-carrier may be variable. Optimization of this bit assignment is the subject of bit allocation, to be discussed in the next chapter.

If the number of sub-carriers is not a power of two, then it is common to add symbols of value zero to the input block so that the advantage of using such a block length in the FFT is achieved. The quantity N which determines the output symbol rate is then that of the padded input block rather than the number of sub-carriers.

The analog filters at the transmitter output and the receiver input should bandlimit the respective signals to a bandwidth of $1/T$. Low pass filters could be used instead of the band-pass filters shown, placed on the other side of the modulator or demodulator. The transmit filter eliminates out-of-band power which may interfere with other signals. The receive filter is essential to avoid aliasing effects. The transmitted signal has a continuous spectrum whose samples at frequencies spaced $1/T$ apart agree with the mapped input data. In particular, the spectrum of each sub-carrier is the form $sinc(1/T)$, whose central value is that input value, and whose nulls occur at the central frequencies of all other sub-carriers.

The receiver operations are essentially the reverse of those in the transmitter. A critical set of functions, however, are synchronization of carrier frequency, sampling rate, and block framing. There is a minimum delay of $2T$ through the system because of the block assembly functions in the transmitter and receiver.

The FFT functions may be performed either by a general purpose DSP or special circuitry, depending primarily on the information rate to be carried. Some simplification compared with full multiplication may be possible at the transmitter, by taking advantage of small constellation sizes.

The number of operations per block of duration T is $KN \log_2 N$, where K is a small quantity. To compare this with a single carrier system, the number of operations per line symbol interval T/N is $K \log_2 N$, which is substantially below the requirement of an equalizer in a typical single carrier implementation for wireline applications.

In most wireline systems it is desirable to transmit the transformed symbols d_n without any further modulation stages. In this case, it is only possible to transmit real line symbols, and not the above complex quantities. The problem is solved by augmenting the original sequence D_n by appending its complex conjugate to it, as shown in Figure 2.3. The *2N*-point IFFT of this augmented sequence is then a sequence of *2N* real numbers, which is equivalent in bandwidth to *N* complex numbers.

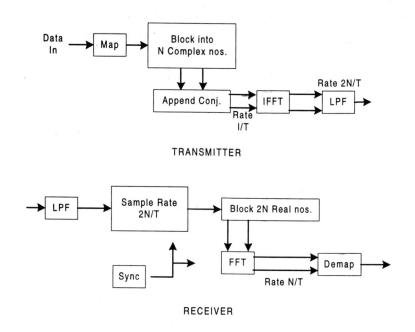

Figure 2.3. System with real transmission.

The augmented sequence is formed from the original sequence as

$$D_n^{'} = \begin{cases} D_n, & n = 1,...,N-1 \\ D_{2N-n}^{*}, & n = N+1,...,2N-1 \end{cases}. \qquad 2.15$$

In order to maintain conjugate symmetry, it is essential that $D_0^{'}$ and $D_N^{'}$ be real. If the original D_0 is zero, as is common, then $D_0^{'}$ and $D_N^{'}$ are set to zero. Otherwise, $D_0^{'}$ may be set to $\text{Re}(D_0)$ and $D_N^{'}$ to $\text{Im}(D_0)$.

For the simple case of $D_0 = 0$, the output of the IFFT is:

$$d_m = \sum_{n=0}^{2N-1} D_n^{'} e^{\frac{j\pi mn}{N}} = 2\,\text{Re}\sum_{n=0}^{N-1} D_n e^{\frac{j\pi mn}{N}} =$$

$$\hspace{6cm} 2.16$$

$$2\sum_{n=0}^{N-1}[A_n \cos\frac{\pi mn}{N} - B_n \sin\frac{\pi mn}{N}], \; m = 0,...2N-1,$$

where $D_n = A_n + jB_n$. Of course the scaling by a factor of two is immaterial and can be dropped. This real orthogonal transformation is fully equivalent to the complex one, and all subsequent analyses are applicable.

2.3   Partial FFT

In some applications, the receiver makes use of a subset of transmitted carriers. For example, in a digital broadcasting system receiver [2] that decodes a group of sub-carriers (channels) or in some multi-rate applications, the receiver has several fall-back modes so using a repetitive structure is advantageous. One of the benefits of an OFDM system with an FFT structure is the fact that it lends itself to a repetitive structure very well. This structure is preferred compared to the required filtering complexity in other wideband systems. Two common structures are shown in Figure 2.4.

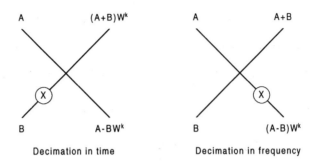

Figure 2.4. Two different techniques for FFT butterfly.

An example of partial FFT is shown in Figure 2.5. In order to detect (receive) the marked point at the output, we can restrict the FFT calculation to the marked lines. Therefore, a significant amount of processing will be saved.

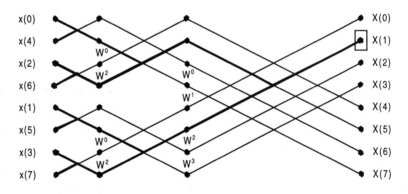

Figure 2.5. Partial FFT (DIT)

Two main differences between decimation in time (DIT) and decimation in frequency (DIF) are noted [1]. First, for DIT, the input is bit-reversed and output is in natural order, while in DIF the reverse is true. Secondly, for DIT complex multiplication is performed before the add-subtract operation, while in DIF the order is reversed. While complexity of the two structures is similar in typical DFT, this is not the case for partial FFT [2]. The reason is that in the DIT version of partial FFT, a sign change (multiplication by 1 and -1) occurs at the first stages, but in the DIF version it occurs in later stages.

2.4 Cyclic Extension

Transmission of data in the frequency domain using an FFT, as a computationally efficient orthogonal linear transformation, results in robustness against ISI in the time domain. Unlike the Fourier Transform (FT), the DFT (or FFT) of the circular convolution of two signals is equal to the product of their DFT's (FFT).

$$FT\{d_n * h_n\} = FT\{d_n\} \times FT\{h_n\}$$
$$DFT\{d_n \otimes h_n\} = DFT\{d_n\} \times DFT\{h_n\},$$

2.17

where * and $\otimes$ denote linear and circular convolution respectively.

Signal and channel, however, are linearly convolved. After adding prefix and postfix extensions to each block, linear convolution is equivalent to a circular convolution as shown in Figure 2.6. Instead of adding prefix an postfix, some systems use only prefix, then by adjusting the window position at the receiver proper cyclic effect will be achieved.

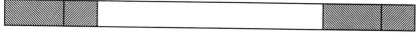

Figure 2.6. Prefix and postfix cyclic extension.

Using this technique, a signal, otherwise aliased, appears infinitely periodic to the channel. Let's assume the channel response is spread over M samples, and the data block has N samples then:

$$y(n) = \sum_{m=0}^{N-1} d(m)h(n-m) \times R_N(n) \qquad n = 0,1, \ldots N+M-1, \qquad 2.18$$

where $R_N(n)$ is a rectangular window of length N. To describe the effect of distortion, we proceed with the Fourier Transform noting that convolution is linear

$$Y(k) = Y(e^{j\omega}) \Big|_{\omega = k\omega_s} = \left\{ D(e^{j\omega}) \times H(e^{j\omega}) \right\} * \frac{\sin \omega \frac{N}{2}}{\sin \frac{\omega}{2}} e^{j\omega \frac{N-1}{2}} \Bigg|_{\omega = k\omega_s}$$

$$2.19$$

After linear convolution of the signal and channel impulse response, the received sequence is of length $N+M-1$. The sequence is truncated to N samples and transformed to the frequency domain, which is equivalent to convolution and truncation. However, in the case of cyclic pre/postfix extension, the linear convolution is the same as the circular convolution as long as channel spread is shorter than guard interval. After truncation, the DFT can be applied, resulting in a sequence of length N because the circular convolution of the two sequences has period of N.

Intuitively, an N-point DFT of a sequence corresponds to a Fourier series of the periodic extension of the sequence with a period of N. So, in the case of no cyclic extension we have

$$\sum_{i=-\infty}^{\infty} \sum_{m=0}^{N-1} d(m)h(n+iN-m), \qquad 2.20$$

which is equivalent to repeating a block of length $N+M-1$ with period N. This results in aliasing or inter-symbol interference between adjacent OFDM symbols. In other words, the samples close to the boundaries of each symbol experience considerable distortion, and with longer delay spread, more samples will be affected. Using cyclic extension, the convolution changes to a circular operation. Circular convolution of two signals of length N is a sequence of length N so the inter-block interference issue is resolved.

Proper windowing of OFDM blocks, as shown later, is important to mitigate the effect of frequency offset and to control transmitted signal spectrum. However, windowing should be implemented after cyclic extension of the frame, so that the windowed frame is not cyclically extended. A solution to this problem is to extend each frame to $2N$ points at the receiver and implement a $2N$ FFT. Practically, it requires a $2N$ IFFT block at the transmitter, and $2N$ FFT at the receiver. However, by using partial FFT techniques, we can reduce the computation by calculating only the required frequency bins.

If windowing was not required, we could have simply used zero padded pre/postfix, and before the DFT at the receiver copy the beginning and end of frame as prefix and postfix. This creates the same effect of cyclic extension with the advantage of reducing transmit power and causing less ISI.

The relative length of cyclic extension depends on the ratio of the channel delay spread to the OFDM symbol duration.

2.5 Channel Estimation

Channel estimation inverts the effect of non-selective fading on each sub-carrier. Usually, OFDM systems provide pilot signals for channel estimation. In the case of time-varying channels the pilot signal should be repeated frequently. The spacing between pilot signals in time and frequency depends on coherence time and bandwidth of the channel. We can reduce the pilot signal overhead by using a pilot signal with a maximum distance of less than the time and coherence bandwidths. Then, by using time and frequency interpolation, the impulse response and frequency response of the channel can be calculated.

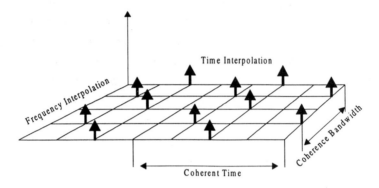

Figure 2.7. Pilot positioning in time and frequency.

Let

$$h_t = (h_1, h_2, \cdots, h_M)^T \qquad\qquad 2.21$$

represent the overall impulse response of the channel, including transmitter and receiver filters, with length M [6]. The N-element training sequence is represented by

$$X = [X_{ij}], \quad 1 \le i \le M + N - 1, \quad 1 \le j \le M$$

$$X_{ij} = \begin{cases} x_{i-j+1}, & 1 \le i - j + 1 \le N \\ 0, & elsewhere \end{cases} . \qquad 2.22$$

The noise samples are

$$n = (n_1, n_2, \cdots, n_{M+N-1})^T, \qquad\qquad 2.23$$

with

$$R_n = E\{nn^*\} = LL^*,$$ 2.24

in which L is a lower triangular matrix, which can be calculated by the Cholesky method. The received signal is then

$$r = Xh_t + n.$$ 2.25

After correlation or matched filtering, the estimate of the impulse response is:

$$\hat{h}_t = X^* r = X^* Xh_t + X^* n.$$ 2.26

If the additive noise is white, the matched filter is the best estimator in terms of maximizing signal-to-noise ratio. However, the receiver filter colors the noise. With a whitening filter the estimate is given in [5]. An unbiased estimator, which removes the effect of sidelobes, is [5]

$$\hat{h}_t = (X^* R_n^{-1} X)^{-1} X^* R_n^{-1} r = h + (X^* R_n^{-1} X)^{-1} X^* R_n^{-1} n.$$ 2.27

In the dual system architecture, circular convolution is replaced by multiplication. The repetition of the training sequence results in circular convolution. Therefore, Equation 2.26 is replaced by:

$$\hat{H}^\omega = X^\omega X^{\omega^*} H^\omega + X^{\omega^*} N^\omega,$$ 2.28

where superscript ω shows Fourier Transform and all products are scalar. The same procedure can be applied to Equation 2.27. This procedure is also called frequency equalization.

2.6   Appendix — Mathematical Modelling of OFDM for Time-Varying Random Channel

Characterization of Randomly Time-Varying Channels

Linear randomly time-varying channels can be characterized by their impulse response , which represents system response at time t due to an impulse at time $t - \tau$. So, for input $x(t)$, the output would be

$$y(t) = \int h(t, \tau)x(t - \tau)d\tau \ .$$

$E\{|h|^2\}$

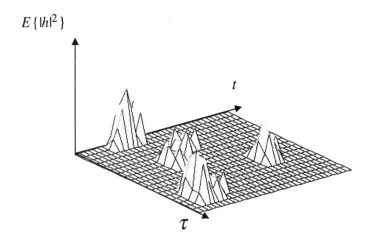

Figure 2.8. Typical impulse response of a wireless channel.

Other kernels can be defined to relate input and output in time and frequency domains [3]. Since the impulse response is time-varying, it is expected that the system transfer function is time-varying. The transfer function is defined as

$$H(t,f) = \int \exp(-j2\pi f\tau)h(t,\tau)d\tau, \qquad 2.29$$

$$y(t) = \int X(f)H(t,f)e^{j2\pi ft}df. \qquad 2.30$$

If the input is a single tone $\exp(j2\pi mt)$ the output is:

$$\int \exp[j2\pi m(t-\tau)]h(t,\tau)d\tau = \exp(j2\pi mt)H(t,m), \qquad 2.31$$

which shows that the time-varying transfer function has an interpretation similar to time-invariant systems. Namely, the output of the system for a sinusoidal function is the input multiplied by the transfer function. In general, $h(t,\tau)$ is a random process and its exact statistical description requires a multi-dimensional probability distribution. A practical characterization of the channel is given by first and second order moments of system functions. For example, many randomly time-varying physical channels are modelled as wide sense stationary non-correlated scatterers. Therefore, they are white and non-stationary with respect to delay variable τ and stationary with respect to t.

$$E\{h(t,\tau)h^*(t',\tau')\} = K(t-u,\tau)\delta(\tau-\tau'), \qquad 2.32$$

$K(t,\tau)$ is called the tap gain correlation. The scattering function of the channel is defined as

$$S(f,\tau) = \int K(\psi,\tau)\exp(-j2\pi f\psi)d\psi. \qquad 2.33$$

For any fixed τ, the scattering function may be regarded as the complex-valued spectrum of the system. The relationship between different system functions is shown in the Figure 2.9 and Figure 2.10.

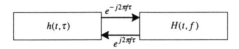

Figure 2.9. Relationship between system functions.

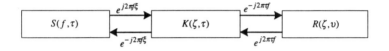

Figure 2.10. Relationship between correlation functions.

Coherence bandwidth of the received signal (or channel) is defined as the frequency distance beyond which frequency responses of the signal are uncorrelated. The reciprocal of delay spread is an estimate for coherence bandwidth. Coherence time is the time distance beyond which the samples of received signal (or channel impulse response) are uncorrelated. Since the channel and signal are Gaussian, independence of samples is equivalent to their being non-correlated. The reciprocal of Doppler spread is an estimate of coherence time [4].

OFDM in Randomly Time-Varying Channels

In general, OFDM is a parallel transmission technique in which N complex symbols modulate N orthogonal waveforms $\psi_i(t)$ which maintain their orthogonality at the output of the channel. This requires that the correlation function of the channel output process, in response to different waveforms, has orthogonal eigenfunctions.

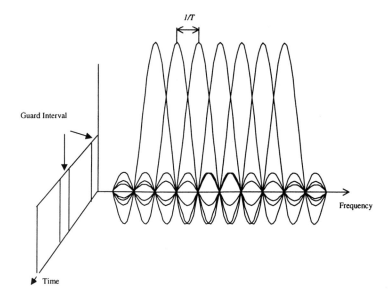

Figure 2.11. OFDM time and frequency span.

Let $h(t,\tau)$ be the Rayleigh distributed, time-varying complex impulse response of the channel, while $R(t,\tau)$ denotes the complex correlation of the output in response to a basic pulse shape $f(t)$:

$$R(t,\tau) = \mathrm{E}\{y(t)y(t+\tau)^*\}.$$ 2.34

Using Karhunen-Loeve expansion, the output of the channel $y(t)$ can be represented as

$$y(t) = \sum_i D_i y_i \varphi_i(t),$$ 2.35

where the $\varphi_i(t)$ s are eigenfunctions of the correlation function $R(t,\tau)$ and

$$y_i = \int y(t)\varphi_i(t)dt,$$ 2.36

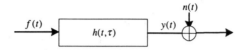

Figure 2.12. Time-varying channel.

Using Karhunen-Loeve expansion, the output of the channel $y(t)$ can be represented as

$$y(t) = \sum_i y_i \varphi_i(t),$$ 2.37

where the $\varphi_i(t)$ s are eigenfunctions of the correlation function $R(t,\tau)$ and

$$y_i = \int y(t)\varphi_i(t)dt.$$ 2.38

The corresponding eigenvalues of the correlation function are λ_i where

$$\int R(t,\tau)\varphi_i(\tau)d\tau = \lambda_i\varphi_i(t).$$ 2.39

For a Gaussian random process $y(t)$, the coefficients in Equation 2.37 are Gaussian random variables and their powers are the corresponding eigenvalues. If the input consists of orthogonal modulating functions which are harmonics of the form

$$\psi_k(t) = u(t)e^{j2\pi f_k t},$$ 2.40

where $u(t)$ is the basic waveform such as time-limted gating function. The eigenfunctions of the received signal correlation function $R_k(t,\tau)$ are,

$$\varphi_i(t)e^{j2\pi f_i t},$$ 2.41

where again $\varphi_i(t)$ are the eigenfunctions' output correlation functions in response to basic waveform $u(t)$. Therefore, the orthogonality condition is

$$\int \varphi_i(t)\varphi_k^*(t)e^{j2\pi(f_i-f_k)t}dt = 0 \qquad\qquad 2.42$$

which can be achieved by increasing $\Delta f = f_{i+1} - f_i$. Analytically this means that l_2 sub-spaces spanned by different eigenfunctions are mutually orthogonal. In other words, the frequency response of the channel for different frequency bins is independent. For a pure tone signal of frequency f_k, the output of random channel is:

$$y_k(t) = \int h(t,\tau)x(t-\tau)d\tau = \int h(t,\tau)e^{j2\pi f_k(t-\tau)}d\tau \qquad\qquad 2.43$$

$$= e^{j2\pi f_k t}\int h(t,\tau)e^{-j2\pi f_k\tau}d\tau$$

The transfer function of a time-varying channel is defined as

$$H(f,t) = \int e^{-j2\pi f\tau}h(t,\tau)d\tau, \qquad\qquad 2.44$$

so the output of channel can be represented as

$$y_k(t) = H(f_k,t)e^{j2\pi f_k t}, \qquad\qquad 2.45$$

and the orthogonality requirement of the output processes holds. If the filter gains $H(f_k,t)$ are orthogonal, and the tap filters are non-correlated and Gaussian, then they will be independent too. As an example, consider orthogonal functions of the form

$$\varphi_{ik}(t) = u(t-iT)e^{2j\pi f_k(t-jT)}. \qquad\qquad 2.46$$

This set of orthogonal functions is used to modulate data symbols and transmit them in a single OFDM symbol timing interval such that the orthogonality in the frequency domain is preserved. By choosing the above set of orthogonal functions, the eigenfunctions used in Karhunen-Loeve expansion will be related as shown in Equation 2.41. The optimum receiver maximizes *a posteriori* probability (conditional probability of the received signal given a particular sequence is transmitted). Usually, channel characteristics do not change during one or a few symbols. Therefore, the channel output can be represented as

$$y(t) = \sum_k H(f_k, t) e^{j2\pi f_k t}.$$

2.47

The received signal $r(t) = y(t) + n(t)$ with $n(t)$ representing additive white Gaussian noise can be expanded as

$$r(t) = \sum_i r_i \varphi_i(t),$$

2.48

where

$$r_i = y_i + n_i$$
$$n_i = \int n(t) \varphi_i(t) dt$$
$$y_i = \int y(t) \varphi_i(t) dt.$$

2.49

Notice that the noise projection should be interpreted as a stochastic integral, because a Riemann integral is not defined for white noise. $r_i s$ are independent zero mean Gaussian random variables and their powers are

$$E|r_i|^2 = E|H_i|^2 + \overline{n_i^2},$$

2.50

assuming that data symbols are of equal and unit power. So, the problem at hand reduces to the detection of a Gaussian random process corrupted by Gaussian noise.

References

1.   Rabiner, L.R., Gold B. *Theory and Application of Digital Signal Processing*. Englewood Cliffs, NJ: Prentice Hall, 1975.

2. Alard, M., Lassalle R. "Principles of Modulation and Channel Coding for Digital Broadcasting for Mobile Receivers." *EBU Review;* Aug 1987; 224: 47-68

3. Bello, P.A. "Characterization of Randomly Time-Variant Linear Channels", *IEEE Trans Comm*; Dec 1963; COM-11; 360-393.

4. Kennedy, R.S. *Fading Dispersive Communication Channels.* New York: Wiley Interscience, 1969.

5. Bahai, A.R.S. "Estimation in Randomly Time-Varying Systems with Applications to Digital Communications." *Ph.D. Dissertation, Univ. of California at Berkeley*, 1993.

6. Klein, A., Mohr, W. "Measurement-Based Parameter Adaptation of Wideband Spatial Mobile Radio Channel Models." IEEE 4th International Symp. on Spread Spectrum Techniques and App. Proc., ISSSTA'95, 91-97

Chapter 3 *Performance Over Time-Invariant Channels*

3.1 Time-Invariant Non-Flat Channel with Colored Noise

Many channels, such as wireline, have transmission properties and noise statistics that vary very slowly with time. Therefore, over moderately long time intervals they may be treated as being invariant. In such a case, the channel impulse response $h(t, \tau)$ is a function only of $t - \tau$, and may be written as $h(u)$ where $u = t - \tau$. Since it is a function of only one variable, we may take its 1-dimensional Fourier Transform to arrive at the channel transfer function

$$H(f) = \int_{-\infty}^{\infty} h(u)e^{-j2\pi f u}du. \qquad 3.1$$

Note that we did not constrain $h(u)$ to be causal. A common practice is to shift the time axis so that $t = 0$ coincides with some principal sampling point of $h(u)$.

Similarly the auto-correlation of the noise $R_N(t, \tau)$ is also a function only of $t - \tau$, and can be treated as $R_n(u)$. Its Fourier Transform $N(f)$ is the power spectral density. Since, in general, both $|H(f)|$ and $N(f)$ vary over the frequency band of interest, the signal-to-noise ratio also varies.

We are now ready to analyze the performance of a multi-carrier system over such a channel, and to optimize the transmitted signal so as to maximize performance.

3.2   Error Probability

Since a fully equalized OFDM system can be treated as N independent QAM signals with independent noise, error probability analysis is the same as for QAM [1]. The QAM signal with independent noise in turn consists of two orthogonal components, so we may deal with each of these components separately.

Each of the I and Q components can be considered to be an independent PAM signal. As noted previously, the signal points in a symmetric L-level PAM signal after demodulation can be written as

$$\{-A(L-1), \cdots, -A, A, \cdots, A(L-1)\}. \tag{3.2}$$

The average power of either the I or Q component of the jth sub-channel is then

$$P = 2A^2 \sum_k p(k)(2k-1)^2, \tag{3.3}$$

where we have used the symmetry of the signal, and $p(k)$ is the probability of level k.

If the levels are equiprobable, as in an L^2-point QAM constellation, then

$$P = \frac{L^2 - 1}{3} A^2.$$ (3.4)

If the noise can be treated as white with power spectral density $N_{0j}/2$ over the sub-channel of interest, then its variance will be

$$N_j = N_{0j}/2T$$ (3.5)

in the I or Q component.

An error will result if the noise amplitude D, after demodulation, is such that $D > A$ or $D < -A$ for an inner level, $D < -A$ for the most positive level, and $D > A$ for the most negative level.

So, the error probability is

$$P_{ej} = 2 \frac{L-1}{L} Q(\frac{A}{\sqrt{N_j}}),$$ (3.6)

where

$$Q(x) = \frac{1}{\sqrt{2\pi}} \int_x^\infty e^{-x^2/2}.$$ (3.7)

It should be noted that for low error probabilities $(x > 3)$, $Q(x)$ may be approximated by

$$Q(x) \approx \frac{1}{x\sqrt{2\pi}} e^{-x^2/2}.$$ (3.8)

Equation 3.6 is the probability of error per I or Q sub-symbol. We are usually concerned with the bit error probability. If the assignment of bits to levels is such that adjacent levels differ by only one bit, as in a Gray code, then the bit error probability becomes

$$P_{ebj} = \frac{P_{ej}}{\log_2 L}.$$ 3.9

Most radio systems use QPSK (a 4-point constellation) on each sub-carrier, so that $L = 2$ and

$$P_{ebj} = Q(\frac{A}{\sqrt{N_j}}).$$ 3.10

For the more general equiprobable case, we substitute

$$P_{ebj} = 2\frac{L-1}{L\log_2 L}Q\left(\sqrt{\frac{3P_j}{N_j(L^2-1)}}\right).$$ 3.11

We can often neglect the coefficient, which is always ≤ 1. Particularly for low error probability and a constellation size not too large, this provides a reasonably tight upper bound.

The quantity P_j / N_j is the signal-to-noise power ratio, where that noise power is measured in a bandwidth equal to the sub-carrier QAM symbol rate. It is also equal to E / N_{0j}, where E is the energy per I or Q sub-symbol. A common representation of the signal-to-noise ratio is E_b / N_0 where E_b is the energy per bit, and

$$E_b = \frac{E}{\log_2 L}.$$ 3.12

The above analysis assumes that the noise power spectral density is constant at $N(f) = N_{0j}/2$ for each sub-channel. For multi-carrier systems with a large number of sub-channels, this should be an excellent approximation. When this is not the case, we should use

$$N_j = \int N(f)R_j^2(f)df \qquad 3.13$$

where $R_j(f)$ is the receiver response for the jth sub-channel.

For straight DMT without windowing,

$$R_j(f) = \frac{\sin\pi(|f|-f_j)T}{\pi(|f|-f_j)T}, \qquad 3.14$$

where f_j is the center frequency of the sub-channel.

It is common in analyzing multi-carrier systems to deal with QAM sub-carriers rather than the I and Q components. Equation 3.11 can be written as

$$p_{ebj} = \frac{K}{b_j}Q\left(\sqrt{\frac{3P_j}{N_j(M_j-1)}}\right), \qquad 3.15$$

where M_j is the number of points in the QAM constellation, P_j is its power, N_j is the noise power in the that sub-channel after receiver processing, and K is the average number of its nearest neighbors, and is usually (but not necessarily) an integer.

Equation 3.15 is strictly applicable to square equiprobable constellations. However, very little error will result if we instead use the same analysis for more general equiprobable constellations. For example, a large QAM constellation with a circular boundary will have only 0.2 dB less average power than a square constellation with the same number of points and spacing between points. If further "shaping gain" is provided by using a multi-dimensional hyperspheric constellation boundary over several QAM sub-channels, then up to 1.53 dB average power reduction can be achieved. However, such shaping gain is rarely used in OFDM systems.

$$P_{eb} = \frac{1}{\sum_j b_j} \sum_j b_j P_{ebj}.$$

3.16

For a complete OFDM system, the overall average bit error probability is

$$P_{ej} = 2\frac{L-1}{L}Q(\frac{A}{\sqrt{N_j}}),$$

3.17

Clearly if the P_{ebj} are all equal, that will also be the overall error probability. Otherwise, the higher error probabilities will dominate performance. A sensible design goal is, therefore, to try to achieve the same error probabilities on all sub-channels if possible.

3.3   Bit Allocation

We now address the problem of optimizing the performance of an OFDM system over a stationary, non-flat, linear channel through choice of the transmitted signal. We may either seek to maximize the overall bit rate with a required error probability, or we may minimize the error probability for a given bit rate. We will first deal with the former optimization, that is, the maximization of

$$R_b = \frac{1}{T}\sum_j b_j,$$

3.18

where b_j is the number of bits carried by the constellation of the jth subcarrier. We will deal primarily with the case of high signal-to-noise ratio (low error probability).

The total transmitted power is assumed to be constrained. We assume that the multi-carrier signal consists of a large number of non-interfering sub-channels, separated in frequency by $\Delta f = 1 / T$. Δf is assumed to be small enough so that the channel gain $H_j(f)$ and the noise power spectral density are constant over each of the sub-channels. So, the channel gain and noise power on the jth sub-channel can be given by H_j and N_j respectively. It is assumed that the noise is Gaussian. This assumption may seem questionable when the noise is primarily due to crosstalk, as on most cabled wire-pair media. However the assumption becomes valid when there are many such interferers, and/or when they are passed through a narrow filter, as is the case in OFDM.

The variables in the optimization are the individually transmitted sub-channel powers P_j and the individual constellation sizes M_j. Fortunately these optimizations are separable. We may first optimize the power distribution, under the constraint

$$\sum_j P_j = P. \qquad 3.19$$

It would appear that the constraint of an overall error probability is best met by requiring that same error probability for each of the sub-channels. This is indeed true at low error probabilities, which is the condition under study here, but it has recently been shown [2] that this is, in general, sub-optimum. A further simplification that is usually made is to constrain the symbol error probability of each sub-channel to the target value, rather than the bit error probability. This can be significantly pessimistic, but the difference in required signal-to-noise ratio is not too great at very low error probabilities, or when trellis coding is used. When fine precision is desired, the difference between symbol and bit error probability can be accounted for.

Setting the symbol error probability to the target value p requires that

$$KQ\left(\sqrt{\frac{3P_j|H_j|^2}{N_j(M_j-1)}}\right) = p. \qquad 3.20$$

This requires that we set the quantity

$$\Gamma^2 = \frac{3P_j|H_j|^2}{N_j(M_j-1)} \qquad 3.21$$

to the constant value such that

$$Q(\Gamma) = \frac{p}{4}. \qquad 3.22$$

The problem of optimizing the transmit powers of the sub-carriers is similar to a classical problem in information theory: given a linear channel with transfer function $H(f)$, Gaussian noise of power spectral density $N(f)$, and a transmit power constraint P, find the transmit power distribution that maximizes the capacity of the channel. The answer is the well-known "water pouring" solution [3]:

$$P(f) = \begin{cases} \lambda - \dfrac{N(f)}{|H(f)|^2}, & f \in F \\ 0, & f \notin F \end{cases}, \qquad 3.23$$

where

$$F = \{f: \frac{N(f)}{|H(f)|^2} \le \lambda\}, \qquad 3.24$$

and λ is the value for which

$$\int_F P(f)df = P. \qquad 3.25$$

This is illustrated in Figure 3.1. The curve of $N(f)/||H(f)||^2$ may be thought of as a bowl into which water whose area is equal to the allotted power is poured. The level of water is λ, and the distance between that level and the curve is the optimum transmit power spectrum. The range of F has been shown as contiguous for simplicity, but, in general, this need not be true.

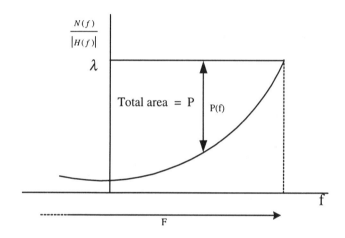

Figure 3.1. Optimum transmit power distribution by the water pouring theorem.

It was shown in [4] that the optimum multi-carrier spectrum, under the assumptions given above, has a very similar form:

$$P_j = \begin{cases} \lambda - \dfrac{N_j \Gamma^2}{3P|H_j|^2} & , \ j \in F \\[4mm] 0 & , \ j \notin F \end{cases},$$ (3.26)

where

$$F = \{j: \frac{N_j \Gamma^2}{3P|H_j|^2} \le \lambda\},$$ 3.27

and λ is the value that causes

$$\sum_j P_j = P.$$ 3.28

Note that over a region for which $N_j / |H_j|^2$ is small, P_j is approximately constant $\approx \lambda$. In a practical system with a low required error probability, we delete the region where $N_j / |H_j|^2$ is not low enough to support a sub-carrier with at least 4 constellation points at the required error probability. The result is an approximately flat spectrum over a region somewhat narrower than F. It was shown in [4] and verified in [5] that using a flat transmit spectrum results in very little degradation over the optimum when the required error probability is low.

A near-optimum design of a multi-carrier system therefore is to set all sub-carriers which cannot support a minimum size constellation to zero, and then, at least initially, to divide the allotted transmit power equally among the remaining sub-carriers. The next problem is to choose the constellation sizes of the non-zero sub-carriers.

Firstly, we will treat the constellation sizes as being continuous variables. Restricting M_j to an integer, which of course is essential, has almost no effect on the result. On the other hand, restricting $b_j = \log_2 M_j$ to an integer, which is convenient but not essential, does have significant effect. Solving Equation 3.21,

$$M_j = 1 + \frac{3P_j|H_j|^2}{N_j \Gamma^2}.$$ 3.29

Rounding down to the next lower integer has little effect, and we have now found an approximation to the problem of maximizing the overall bit rate under an error probability constraint. The resultant bit rate is

$$R_b = \frac{1}{T} \sum_j \log_2 [1 + \frac{3P_j |H_j|^2}{N_j \Gamma^2}].$$

3.30

Since the sub-channels are closely spaced, the above may be closely approximated by the integral

$$R_b = \int_F \log_2 [1 + \frac{3P(f)|H_j|^2}{N(f)\Gamma^2}] df.$$

3.31

The above is the same as the well-known formula for the capacity of the channel, in which the signal-to-noise ratio,

$$\rho(f) = \frac{P(f)|H_j|^2}{N(f)}$$

3.32

has been decreased by a factor of $\frac{3}{\Gamma^2}$. This factor has been referred to as the "gap" [6]. In other words, the actual rate achievable, at a given error probability, on a given channel by an uncoded multi-carrier system is equal to the capacity of the same channel in which the signal-to-noise ratio has been reduced by the gap.

The above integral (Equation 3.31), except for a small change in the range of integration, is the same as the rate achievable over the same channel by a single carrier system using an ideal decision feedback equalizer (DFE). This leads to the very interesting result that at low error probability, the performance of an optimized multi-carrier system is approximately the same as that of a single carrier system with an ideal DFE [8]. This is particularly true when compared with an MMSE DFE as opposed to a zero-forcing DFE

[9]. At higher error probabilities, the multi-carrier system has been shown to perform somewhat better [2].

We can rewrite Equation 3.30 as:

$$R_b = \frac{1}{T}\sum_j \log_2[1 + \frac{3\rho_j}{\Gamma^2}] = \frac{1}{T}\log_2\{\prod_j [1 + \frac{3\rho_j}{\Gamma^2}]\}. \qquad 3.33$$

Now define an average signal-to noise ratio $\overline{\rho}$ [6] such that

$$R_b = \frac{1}{T}\log_2[1 + \frac{3\overline{\rho}}{\Gamma^2}], \qquad 3.34$$

Then, with n the number of sub-channels,

$$\overline{\rho} = \frac{\Gamma^2}{3}\{(\prod_{j=1}^{n} [1 + \frac{3\rho_j}{\Gamma^2}])^{\frac{1}{n}} - 1\}. \qquad 3.35$$

and at high signal-to-noise ratio,

$$\overline{\rho} = [\prod_{j=1}^{n} \rho_j]^{\frac{1}{n}}. \qquad 3.36$$

Thus, at high signal-to-noise ratio, when a multi-carrier system is fully optimized, the average signal-to-noise ratio is the geometric mean of the signal-to-noise ratios of the sub-channels. This is an important quantity, and serves as a good quality measure of the channel. The integral form of the above appears in analyses of single carrier systems with DFE.

So far we have not required that the number of bits carried by each sub-carrier be an integer. Such non-integral values can be achieved by treating more than one sub-carrier as a constellation in a higher-dimensional space, and assigning an integral number of bits to that higher-dimensional constellation. However, to avoid that complexity, it is usually required that b_j · be an integer. Simply performing the above optimization and rounding each b_j down results in the average loss of ½ bit per sub-carrier.

The above loss can be mostly recovered by small adjustments in the sub-carrier powers. Those sub-carriers that required a large round-down can have their powers increased slightly in order to achieve one more bit. Those with a smaller round-down can have their power reduced while keeping the same number of bits. During this process of power re-allocation, it is important to ensure that the constraint of total power is kept. After the process is completed, the power deviations will vary over a 3 dB range.

A further small increase in bit rate can be achieved by accounting for the bit error probability of each sub-carrier rather than the symbol error probability, which is higher. This involves calculating the bit error probability for each sub-carrier after the system has been optimized for symbol error probability. Those sub-carriers which can meet the bit error probability target with an addition bit can then have their constellations increased accordingly.

3.4 Bit and Power Allocation Algorithms for Fixed Bit Rate

A more common design requirement is to minimize the bit error probability for a fixed bit rate, where that error probability must be below some target value. Again, we will assume a transmitter power constraint, and integral number of bits per sub-channel. Several algorithms have been developed to solve this design problem, with varying degrees of precision, complexity, and computation time [9].

One approach is similar to the previously described procedure for maximizing a variable bit rate. As before, we first determine Γ such that $Q(\Gamma) = p / 4$, where p is the desired error probability, and is conservatively chosen as the maximum symbol error probability for each sub-carrier. We now examine if the unquantized total bit rate

$$R_b = \frac{1}{T}\sum_j \log_2[1 + \frac{3P_j|H_j|^2}{N_j\Gamma^2}]\qquad 3.37$$

is greater than the required rate. If not, the desired performance cannot be achieved. Otherwise, if $R_b - B \geq \frac{A}{T}J$, where B is the required bit rate, A is a positive integer, and J is the number of used sub-carriers, then A bits are subtracted from each sub-carrier. This step adds 3 dB of noise margin.

After the above procedure, the number of bits on each sub-carrier is rounded down to the next lowest integer. The resultant bit rate R' is compared with B. If $R' > B$, then a set of sub-carriers with the smallest round-off are each reduced by an additional bit. If $R' < B$, then a set of sub-carriers with the largest round-off are each increased by one bit. Finally, the powers of the sub-carriers are adjusted as before to achieve the same error probability for each.

References

1.   Gitlin, R.D., Hayes J.F., Weinstein S.B. *Data Communications Principles*. New York: Plenum, 1992.

2.   Willink, T.J., Wittke P.H. "Optimization and Performance Evaluation of Multi-Carrier Transmission." *IEEE Trans. Info. Theory*; Mar 1997; 43: 426-440.

3.   Gallagher, R.G. *Information Theory and Reliable Communication*. New York: Wiley, 1968.

4.   Kalet, I. "The Multitone Channel." *IEEE Trans. Commun.*; Feb 1989; 37: 119-124.

5.   Feig, E. "Practical Aspects of DFT-Based Frequency Division Multiplexing for Data Transmission." *IEEE Trans. Commun.*; Jul 1990; 38: 929-932.

6. Sistanizadeh, K., Chow P.S., Cioffi J.M. "Multi-Tone Transmission for ADSL." *IEEE Int. Conf. Commun.*; 1993; 756-760.

7. Chow, P.S., Cioffi J.M., Bingham J.A.C. "A Practical Discrete Multitone Transceiver Loading Algorithm for Data Transmission Over Spectrally Shaped Channels." *IEEE Trans. Commun.*, Feb-Apr 1995; 43: 773-775.

8. Zervos, N.A., Kalet I. "Optimized DFE Versus Optimized OFDM for High-Speed Data Transmission Over the Local Cable Network." *IEEE Int. Conf. Commun.*; 1993; 35.2.

9. Cioffi, J.M., Dudevoir GP, Eyuboglu M.V., Forney G.D. "MMSE Decision Feedback Equalizers and Coding — Part I: Equalization Results." *IEEE Trans. Commun.*, Oct 1995; 43: 2582-2594.

10. Hughes-Hartog, D. *U.S. Patent 4,731,816; U.S. Patent 4,833,706.*

11. Chow, P.S., Cioffi J.M., Bingham J.A.C. "A Practical Discrete Multitone Receiver Loading Algorithm for Data Transmission Over Spectrally Shaped Channels." *IEEE Trans. Commun.; Feb 1995; 43: 773-775.*

Chapter 4 *Clipping in Multi-Carrier Systems*

4.1 Introduction

It is widely recognized that a serious problem in OFDM is the possibility of extreme amplitude excursions of the signal. The signal is the sum of N independent (but not necessarily identically distributed) complex random variables, each of which may be thought of as a Quadrature Amplitude Modulated (QAM) signal of a different carrier frequency. In the most extreme case, the different carriers may all line up in phase at some instant in time, and therefore produce an amplitude peak equal to the sum of the amplitudes of the individual carriers. This occurs with extremely low probability for large N.

The problem of high peak amplitude excursions is most severe at the transmitter output. In order to transmit these peaks without clipping, not only must the D/A converter have enough bits to accommodate the peaks, but more importantly the power amplifier must remain linear over an amplitude range that includes the peak amplitudes. This leads to both high cost and high power consumption.

Several researchers have proposed schemes for reducing peak amplitude by introducing redundancy in the set of transmitted symbols so as to eliminate those combinations which produce large peaks [1, 2]. Clearly the required redundancy increases with the desired reduction in peak amplitude. This redundancy is in addition to any other coding used to improve performance, and leads to a reduction in the carried bit rate.

In this chapter, we will examine the effects of clipping an unconstrained OFDM signal. Several previous papers have analyzed the clipping as an additive Gaussian noise. This approach is reasonable if the clipping level is sufficiently low to produce several clipping events during an OFDM symbol interval. In most realistic cases, however, particularly when the desired error probability is low, the clipping level is set high enough such that clipping is a rare event, occurring substantially less than once per OFDM symbol duration. Clipping under these conditions is a form of impulsive noise rather than a continual background noise, leading to a very different type of error mechanism. Here we evaluate the rate of clipping with a high clipping threshold, and the energy and spectrum of those events. This permits determination of error probability and interference into adjacent channels.

The transmitter output of a multi-carrier system is a linear combination of complex independent random variables. It is generally assumed that the distribution of that output signal is Gaussian using the central limit theorem. This assumption is valid for large N over the range of interest.

The output of the transmitter Inverse Fourier Transform, $\{d_m\}$ is a linear combination of complex independent identically distributed symbols $\{D_n\}$ of the corresponding constellations.

$$d_m = \frac{1}{\sqrt{N}} \sum_{n=0}^{N-1} D_n e^{j2\pi n \frac{m}{N}}.$$

                                                                        4.1

Assuming Gaussian distribution for transmitted samples, signal "peak" should be defined carefully. Peak-to-average has commonly been used, yet a clear definition of peak is not explicitly presented. Absolute peak is not a proper parameter in our analysis because the probability of co-phasing N complex independent random variables is very small, even for moderate values of N. A proper definition should take into account statistical characteristics of the signal and the fact that Gaussian distribution tails are extended infinitely. A reasonable alternative to absolute maximum is to define peak of signal power (variance) such that probability of crossing that level is negligible. For example, a threshold of 5.2 times the average rms (5.2σ), corresponds to a crossing probability of 10^{-7}. Such a definition of peak-to-average ratio of a Gaussian signal, unlike the absolute maximum, is independent of the number of sub-carriers.

4.2 Power Amplifier Non-Linearity

Several analytical models for power amplifier non-linearity are offered in the literature [9]. Amplitude and phase non-linearity of power amplifiers known as AM/AM and AM/PM non-linearity results in an inter-modulation distortion and requires operation of amplifiers well below a compression point as shown in Figure 4.1.

For a narrowband Gaussian input signal[2] $x(t)$ of the form:

$$x(t) = I(t)\cos \omega t - Q(t)\sin \omega t = R(t)\cos(\omega t + \varphi(t)), \qquad 4.2$$

---

[2] It means the ratio of signal bandwidth to carrier frequency is small. Majority of wireless modulation systems meet this requirement.

the output of the non-linear device using AM/AM and AM/PM non-linearity model is:

$$y(t) = A(R(t))\cos(\omega t + \varphi(t) + \theta(t)).$$ 4.3

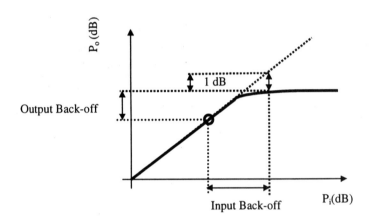

Figure 4.1. Power amplifier 1 dB compression point.

Let $f(.)$ denote the instantaneous mapping of the non-linear function. Then $A(.)$ and $f(.)$ are related through the Chebychev transformation [15]:

$$A(R) = \frac{2}{\pi} \int_0^\pi f(R\cos\varphi)\cos\varphi\, d\varphi.$$ 4.4

A proper clipping model for the signal and its envelope is critical in this analysis. We assume a soft amplitude limiter model for the power amplifier.

The relationship between signal clipping and envelope clipping transfer function according to Equation 4.4 is shown in Figure 4.2.

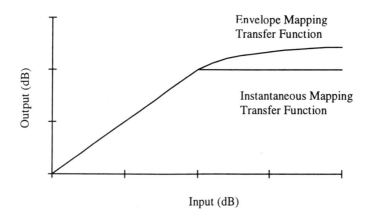

Figure 4.2. Relationship between instantaneous and envelope clipping.

In the first case, the Gaussian signal $x(t)$ is clipped and higher frequencies generated by the non-linearity are filtered out at the output of the amplifier as shown in Figure 4.3. In the latter, the amplitude $R(t)$, which is a Rayleigh process, is clipped and then the output harmonics are filtered out.

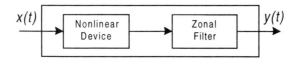

Figure 4.3. A memoryless non-linear device.

Statistics of extreme values for Gaussian and Rayleigh processes are essential for understanding and analysis of clipping in multi-carrier systems. Most of the results in this section apply to both Gaussian and Rayleigh distributed random processes [13]. Assume $x(t)$ is a stationary and continuous with probability one process with finite second order moment. The expected value of number of crossings[3] of level l during time period T by $x(t)$ is [13]

$$T(\frac{m_2}{m_0})^{1/2} e^{-\frac{-l^2}{2m_0}},$$

4.5

where

$$m_i = \int m^i dF(m) \qquad i = 0,2,\cdots$$

4.6

and $dF(m)$ is the power spectral density of $x(t)$. Without loss of generality we assume unity power, $m_0 = 1$. The important quantity m_2 is the power of the derivative of $x(t)$.

Rice and others [8, 10] showed that sequences of up-crossings of an ergodic and stationary process with continuous sample function with probability one, asymptotically approaches a Poisson process. For a *Gaussian signal*, the rate of the Poisson process is

[3] A random process $x(t)$ has an up-crossing of the level at time if

$\exists (\varepsilon > 0); \ (x(t(\leq l) \ \ for \ (t \in (t_0 - \varepsilon, t_0)) \ \ and \ x(t) \geq l \ \ for \ (t \in (t_0, t_0 + \varepsilon))$

$$\lambda_l = \sqrt{m_2} e^{\frac{-l^2}{2}}.$$ 4.7

The shape of the pulse above level l is a parabolic arc of the form

$$P(t) = (-2\pi^2 m_2 t^2 + 2\tau\pi^2 m_2 t + 1)l.$$ 4.8

where τ, duration of clip, is a random variable with Rayleigh probability density

$$P_\tau(\tau) = l^2 \tau (\pi m_2)^2 e^{\frac{-(l\pi m_2)^2 \tau}{2}}.$$ 4.9

The above approximation is valid for $l > 4\sigma^2$.

The envelope of signal $R(t)$ has similar characteristics as a signal $x(t)$ with twice the bandwidth. Therefore, we assume that the shape of excursion above level R of the signal envelope is

$$P_R(t) = (-2\pi^2 m_2 t^2 + 2\tau\pi^2 m_2 t + 1)R.$$ 4.10

The excursion pulse is completely characterized by the parameter τ and second order moments of the signal.

4.3 BER Analysis

Usually, the effect of clipping is modelled as an extra additive noise with a variance equal to the energy of clipped portion. This model does not describe the instantaneous nature of clipping phenomena as a rare event. We define a conditional bit error rate measure to underline the effect of clipping on bit error rate in a multi-carrier system.

Defining event A as clipping of the signal above level l, it is clear that

$$P(symbol\ error) = P(symbol\ error|A)P(A) + P(symbol\ error|\overline{A})P(\overline{A}) \quad 4.11$$

Probability of occurrence of a clipping event during an interval of length T is λ_l and we assume that duration of clipping τ is short compared to the OFDM symbol duration. The spectrum of an OFDM signal asymptotically tends to a rectangular spectrum as the number of carriers increases.

The mean of the corresponding Poisson process is:

$$\lambda_l = \frac{f_0}{\sqrt{3}} e^{-\frac{l^2}{2}}, \quad 4.12$$

where $f_0 = N\Delta f$ for baseband signal. The expected value of the duration of the clip is

$$\overline{\tau} = \frac{1}{2l\sqrt{\pi m_2}} = \frac{1}{2lf_0}\sqrt{\frac{3}{\pi}}. \quad 4.13$$

We use a linear model to show the effect of clipping distortion on the transmitted signal. The clipped signal is represented as $x(t) + C_L(t)$ where the second term includes part of the signal above level l. However, the second term is correlated with the signal [8]. The Gram-Schmidt technique can be used to de-correlate signal and distortion terms. Defining a new distortion term as

$$C_L(t) - \frac{E\{C_L x(t)\}}{E\{x(t)^2\}} x(t),$$

the signal and distortion terms are non-correlated. However, the second term has negligible power compared to the first term. The above decomposition can also be justified by Bussgang's theorem [14]. For a memoryless non-linear function $f(.)$, Bussgang's theorem proves that

$$E\{x(t+\tau)y(t)\} = \gamma E\{x(t+\tau)x(t)\}.\qquad 4.14$$

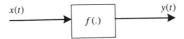

Figure 4.4. Memoryless non-linear mapping.

Therefore, we can decompose the output signal into two non-correlated components:

$$y(t) = \gamma x(t) + C_L(t),\qquad 4.15$$

where $x(t)$ and $c_L(t)$ are "first order" non-correlated .

Using parabolic characteristics of the clipped component, the Fourier Transform of the clipped signal will be

$$g_\tau(\omega) = (2\pi)^2 m_2^2 \frac{l\tau}{\omega^2}(\text{sinc}(\frac{\omega\tau}{2}) - \cos\frac{\omega\tau}{2}).\qquad 4.16$$

This function is shown in Figure 4.5. By using the expected value of distortion power in an arbitrary frequency bin we can estimate the overall distortion effect. The probability distribution function of τ is

$$f(\tau) = (2\pi l m_2)^2 \exp\left(\frac{-(2\pi l m_2 \tau)^2}{8}\right).\qquad 4.17$$

The effect of the noise term in every bin is expected to be increased by

$$\int_{-\infty}^{\infty} |g_\tau(\omega)|^2 \text{sinc}^2(\omega - \omega_k) d\omega.\qquad 4.18$$

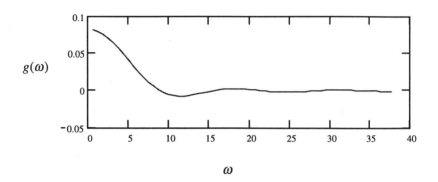

Figure 4.5. Frequency response of clipped signal.

Notice that m_2 is calculated for the entire bandwidth but the integral is over a <u>single</u> bin. We will examine the error probability due to clipping alone, for a baseband clip. It will be assumed that the clip duration is much smaller than the OFDM symbol duration, but larger than the system sampling rate ($\tau \ll T$, $\tau > T/N$). Otherwise the clip is likely to be missed by the signal processing. Of course this implies that N, the number of sub-carriers, is large.

First we will evaluate the effect of a clip on each of the sub-channels. The effect of a clip of duration τ, occurring at time t_o, on the kth sub-channel is

$$F_k = \frac{1}{\sqrt{N}} \sum_{n=0}^{N-1} f_n e^{-j2\pi nk/N} , \qquad\qquad 4.19$$

where f_n are samples of the clip pulse $P(t)$ given by Equation 4.8, and the $\frac{1}{\sqrt{N}}$ factor preserves total power.

$$f_n = P(n\frac{T}{N} - t_0).$$ 4.20

We will replace the discrete Fourier Transform by the conventional continuous one by substituting

$$x_n = \frac{1}{\Delta} \int_{n\Delta - \Delta/2}^{n\Delta + \Delta/2} x(t)dt,$$ 4.21

where $\Delta = \frac{T}{N}$. The above approximation is valid because $\Delta \ll T$. Then

$$F_k = \frac{N}{T}\frac{1}{\sqrt{N}} \int_{t_0 - \tau/2}^{t_0 - \tau/2} P(t - t_0)e^{-j2\pi kt / T}dt.$$ 4.22

Substituting $u = t - t_o$,

$$F_k = \frac{\sqrt{N}}{T}e^{-j2\pi kt_0 / T} \int_{-\tau/2}^{\tau/2} P(u)e^{-j2\pi ku/T}du$$ 4.23

$$F_k = \frac{\sqrt{N}}{T}e^{-j2\pi kt_0/T} g_\tau(\frac{2\pi k}{T}),$$ 4.24

where $g_\tau(\omega)$ is the pulse spectrum given by 4.16, so that

$$F_k = \sqrt{N}Te^{-j2\pi kt_0/T} m_2 \frac{l\tau}{k^2}[\operatorname{sinc}\frac{\pi k\tau}{T} - \cos\frac{\pi k\tau}{T}].$$ 4.25

Since

$$\operatorname{sinc}(x) - \cos(x) \approx -\frac{x^2}{3} \text{ for } x \ll 1,$$ 4.26

the response of the lower sub-carriers are approximately equal, with the higher sub-carriers being progressively reduced.

The error probability will vary over the sub-carriers. However, the overall error probability will be dominated by that of the lower sub-carriers, when the constellation sizes are equal, or even more so when the lower constellation sizes are larger, as is usual in ADSL. For small k, the response to a clip of duration τ, in each sub-carrier, will be

$$F_k = \frac{\pi^2}{3T} \sqrt{N} m_2 \, l \tau^3 \, e^{j\Theta} \qquad\qquad 4.27$$

where Θ is uniformly distributed on $[0, 2\pi]$, and the Rayleigh probability distribution of τ is given by Equation 4.9.

Writing Equation 4.27 as

$$F_k = r \, e^{j\Theta}, \qquad\qquad 4.28$$

we wish to find the probability distribution of r.

$$\Pr[r > R] = \Pr\left[\tau > \left(\frac{3RT}{\pi^2 \sqrt{N} m_2 l}\right)^{1/3}\right]. \qquad\qquad 4.29$$

Using the distribution of τ given by Equation 4.9,

$$\Pr[r > R] = \exp\left\{-\left[\left(\frac{3RT}{\pi^2 \sqrt{N} m_2 l}\right)^{2/3} \pi^2 l^2 m_2\right]\right\}$$

$$= \exp\left\{-\left[\frac{9}{N} R^2 T^2 \pi^2 l^4 m_2\right]^{1/3}\right\}. \qquad\qquad 4.30$$

Substituting

$$m_2 = \frac{N^2}{3T^2}, \qquad\qquad 4.31$$

$$\Pr[r > R] = \exp\left\{-\left[3R^2 \pi^2 N \, l^4\right]^{1/3}\right\}. \qquad\qquad 4.32$$

If we warp the complex plane by mapping $ae^{j\Theta} \rightarrow a^{1/3}e^{j\Theta}$, then Equation 4.32 is a Rayleigh distribution in that warped plane. The real or imaginary component of F_k therefore has a normal distribution with

$$\Pr[r\cos\Theta > x] = Q\left[\frac{x^{1/3}}{\sigma}\right], \qquad\qquad 4.33$$

where

$$Q(x) = \frac{1}{\sqrt{2\pi}}\int_x^\infty e^{-u^2/2}du, \text{ and } \sigma = \left[\sqrt{24N}\pi l^2\right]^{-1/3}. \qquad 4.34$$

We will assume the sub-carrier carries a square constellation of L^2 points. Each component then has L levels, equally spaced and separated by $2d$. The power in that component is $\dfrac{1}{2N}$, since we have normalized the total power to unity. Then

$$d = \sqrt{\frac{3}{2N(L^2-1)}}. \qquad\qquad 4.35$$

The probability of error in the sub-carrier component, given that a clip occurred, is

$$\Pr(e|clip) = \frac{2(L-1)}{L}Q(\frac{d^{1/3}}{\sigma}) = \frac{2(L-1)}{L}Q\left[\left(\frac{6\pi l^2}{\sqrt{L^2-1}}\right)^{1/3}\right]. \qquad 4.36$$

It is interesting to note that the above quantity is independent of N. Although the effect of a clip on a sub-carrier diminishes with the number of sub-carriers, the signal power in each sub-carrier is also inversely proportional to N. Also, not only does the probability of a clip decrease with l, the clipping level, but so does the error probability of error due to each clip. Equation 4.36 is plotted in Figure 4.6 for several values of L. Note that this probability is extremely low for the 4-point constellation ($L=2$), so the effect of a clip is negligible for normalized clip levels above 4, which should almost always be the case. Even for a constellation size of 256 ($L=16$), the probability of a clip producing an error is approximately 10^{-3} when the clip level is 5.2. These error probabilities are upper bounds, that are tight for the lower order sub-carriers.

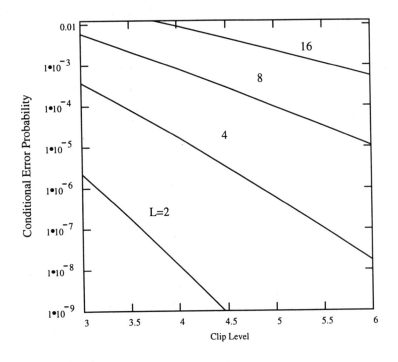

Figure 4.6. Probability of error given a clip.

The quantity of Equation 4.36 must now be multiplied by the probability of a clip. We will assume that clips occur rarely enough so that the probability of more than one clip during an OFDM symbol is negligible.

Then the probability of a clip, either positive or negative, during a symbol interval is

$$\Pr(clip) = 2\lambda_l T = 2\sqrt{m_2}\, T e^{-l^2/2},$$ 4.37

so the probability of error is

$$\Pr(e) = \frac{4\sqrt{3}(L-1)}{L} N e^{-l^2/2} Q\left[\left(\frac{6\pi l^2}{\sqrt{L^2-1}}\right)^{1/3}\right].$$ 4.38

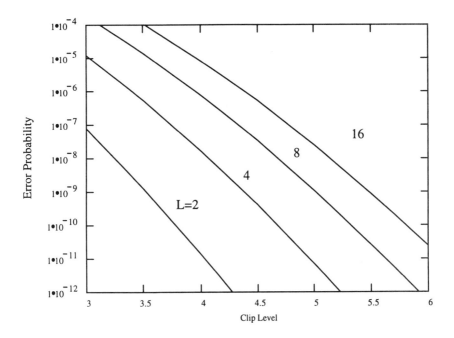

Figure 4.7. Symbol error probability due to clipping.

The quantity $\Pr(e)/N$ is plotted in Figure 4.7. Again it should be stressed that this is a lower bound, tight only for low frequency sub-carriers. It is negligibly small for small constellations, and for moderately high clipping levels.

For the case $L = 2$, which is usual is wireless OFDM systems,

$$\Pr(e) = 2\sqrt{3}N\, e^{-l^2/2} Q\left[\left(2\sqrt{3}\pi l^2\right)^{1/3}\right].$$

4.39

For the case of large constellation size, as occurs in some sub-carriers of an ADSL system,

$$\Pr(e) \approx 4\sqrt{3}N\, e^{-l^2/2} Q\left[\left(\frac{6\pi l^2}{L}\right)^{1/3}\right].$$

4.40

The error probability per bit, assuming Gray coded levels in each sub-carrier component, and noting that errors to other than adjacent levels are negligible, is then

$$\Pr_b(e) \approx \frac{4\sqrt{3}N}{\log_2 L}\, e^{-l^2/2} Q\left[\left(\frac{6\pi l^2}{L}\right)^{1/3}\right].$$

4.41

It should be noted that this value holds in the presence of fading, provided that the fading is slow compared to the OFDM symbol rate, and the clip occurs at the transmitter.

If the clip occurs at the transmitter, and additive Gaussian noise of amplitude n is present on the channel at that time, then the per bit error probability in a component of a lower sub-channel when a clip occurs is

$$\Pr(e|\text{clip, noise}) = \frac{2(L-1)}{L\log_2 L} \int \Pr(r\cos\theta > d - n) \, P(n) \, dn$$

$$= \frac{2(L-1)}{L\log_2 l \sqrt{2\pi}\sigma_n} \int Q\left[\frac{-(d-n)^{1/3}}{\sigma_n}\right] \exp\left[\frac{-n^2}{2\sigma_n}\right] dn, \qquad 4.42$$

where σ_n^2 is the noise power in that component.

The overall per bit error probability is then upper bounded by

$$\Pr(e) < \frac{2(L-1)}{L\log_2 L}\left\{\int \frac{\sqrt{6}Ne^{-l^2/2}}{\sigma_n\sqrt{\pi}} Q\left[\frac{-(d-n)^{1/3}}{\sigma_n}\right] \exp\left[\frac{-n^2}{2\sigma_n}\right] dn + Q\left(\frac{d}{\sigma_n}\right)\right\},$$

4.43

where d is half the distance between constellation points and is given by Equation 4.35. The bound is tight and may be treated as an approximate equality for large L. Of course, if a clip occurs at the receiver, and if fading is present on the channel, the error probability will not be as given above.

Consider a noiseless Rayleigh faded signal with clipping in the receiver. It will be assumed that the clipping is slow compared to the symbol rate. The signal at the clipping point will be normalized to unity power, multiplied in amplitude by a unity mean Rayleigh random variable z, assumed constant over a symbol interval and with longer term probability distribution

$$p(z) = \frac{\pi}{2} z e^{-\frac{\pi z^2}{4}}, \; z > 0. \qquad 4.44$$

For a symbol during which the amplitude variable is z, the error probability is upper bounded by

$$\Pr(e) < \frac{4\sqrt{3}(L-1)}{L} N e^{-l^2 z^2/2} Q\left[\left(\frac{3\pi l^5 z^5}{4\sqrt{L^2-1}}\right)^{1/3}\right]. \qquad 4.45$$

The overall probability of error is then upper bounded by

$$\Pr(e) < \frac{2\pi\sqrt{3}(L-1)}{L} N \int_0^\infty e^{-z^2(\frac{\pi}{4}+\frac{l^2}{2})} Q\left[\left(\frac{3\pi l^5 z^5}{4\sqrt{L^2-1}}\right)^{1/3}\right] dz. \qquad 4.46$$

The error probability $\Pr(e)/N$ from Equation 4.46 is given in Figure 4.8.

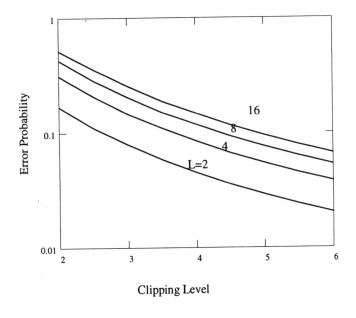

Figure 4.8. Clipping error probability under Rayleigh fading.

It is interesting to compare these results with those that assume that the error due to clipping can be treated as an additive Gaussian noise, an assumption that is valid at low clipping levels where several clips occur during an OFDM symbol interval. In Reference [5], the clipping noise power is given as

$$\sigma_c^2 = -\sqrt{\frac{2}{\pi}} \, l \, e^{-l^2/2} + 2(1+l^2)Q(l), \qquad\qquad 4.47$$

where we have normalized that result to unity signal power. This result is pessimistic in that it does not account for the fact that some of the clipping noise power is out-of-band. A more accurate expression is given in [6].

For the above Gaussian noise level, the per-symbol error probability for a sub-carrier is

$$P_e = \frac{2(L-1)}{L} Q\left(\frac{\sqrt{3}}{\sigma_c \sqrt{L^2-1}} \right). \qquad\qquad 4.48$$

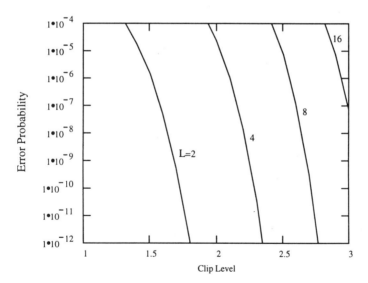

Figure 4.9. Error probability treating clipping as a Gaussian noise.

This is plotted in Figure 4.9 below. Comparing these calculated error probabilities, in particular at a clip level $l = 3$, with those of Figure 4.7, we can see that they are much lower even under pessimistic approximations. This is due to the assumption that the noise resulting from clipping is spread uniformly over time, rather than concentrated as impulses.

4.4 Bandwidth Regrowth

Bandwidth regrowth or out-of-band leakage due to clipping is a critical issue for wireless communications system designers. The amount of out-of-band spill determines filtering requirements and adjacent channel interference. Of course analog filtering is also required at the receiver to reduce aliasing. The spectrum of an instantaneously clipped signal has out-of-band band energy at harmonics of the original fundamental band which reduces the energy of in-band signal. Envelope clipping does not result in any harmonics but the energy spill in the adjacent channel is higher.

Using proper filtering of the power amplifier output mitigates both effects. However, proper choice of filtering and its feasibility is important for overall system performance and cost. Interestingly, filtering of harmonics results in higher fluctuation, and therefore, higher peak-to-average ratio in the received signal.

In this analysis, we use two different approaches to evaluate the impact of clipping on OFDM signals. The first technique is based on calculation of the correlation function of the clipped signal at arbitrary level in order to find the average power spectral density of the leakage. Secondly, we outline how to take into account the impulsive nature of the leakage.

Comparing these two techniques one can verify the validity of asymptotic techniques and provide a basis for a system designer as to how much bandwidth spread can be tolerated.

One should notice that the 'bursty' nature of clipping should be taken into account as in the previous section. Specifically, spectrum estimation requires an averaging, therefore conditional spectral estimation is a more realistic measure of out-of-band leakage. Otherwise, instantaneous power spill is averaged to a smaller value and may mislead system designers.

Following Rice approximation of the clipping function:

$$f(x) = \begin{vmatrix} x, & \text{if } |x| < l \\ \dfrac{|x|}{x}, & \text{elsewhere} \end{vmatrix} \qquad\qquad 4.49$$

by a contour integral of the form:

$$f(x) = \frac{1}{2\pi} \int_C \frac{e^{i(x-l)z} - e^{ixz}}{z^2} dz - \int_{C'} \frac{e^{i(x+l)z} - e^{ixz}}{z^2} dz, \qquad 4.50$$

where C is a path of integration along real axis from $-\infty$ to ∞ and indented downwards near origin and C^1 is the same path but indented upward near origin, the relationship between input correlation function *r(t)* and output correlation function *R(t)* is:

$$\frac{1}{\sqrt{N}} \quad R(t) = r\{erf(l/\sqrt{2})\}^2 + \sum_{n=3,5,\ldots}^{\infty} \frac{r^n}{n!} [H_{n-2}(l)e^{-l^2/2}]^2, \qquad 4.51$$

where $H_n(x)$ are Hermite polynomials. Reference [6] uses this approach to arrive at the power spectrum

$$S(f) = erf^2(\frac{l}{\sqrt{2}})S_{xx}(f) + \sum_{n=2,4,6..}^{\infty} C_n S_{xx}^{(n+1)}(f), \qquad 4.52$$

where $S_{xx}(f)$ is the power spectral density before clipping, $S(f)$ is the spectrum after clipping , $S_{xx}^{(n+1)}(f)$ denotes *(n+1)*-fold convolution, and

$$C_n = \frac{4H_{n-1}^2(l/\sqrt{2})}{\pi 2^n(n+1)!}e^{-l^2}.$$

4.53

We are interested in evaluating the above in the region $f > f_0$, where the unclipped spectrum is zero.

In Reference [8], another expression for the power spectrum is given, including high and low frequency asymptotes. In terms of the notation used in this text, the high frequency asymptotic expression is

$$S(f) = \frac{4\sqrt{6}f_0^3\, e^{-l^2/2}}{9\pi^2 f^4}.$$

4.54

The out-of-band spectrum discussed above must be multiplied by the transmitter filter function $|H_T(f)|^2$. The analysis applies when the clipping level l is low. It is also applicable in any case for determining the long term average power spectral density, for example to examine whether the signal meets regulatory requirements.

In another approach, as shown in the previous section clipping is considered as a rare event, hence approaches a Poisson process. We use the approach outlined in [8] where the random pulse train is

$$r(t) = \sum_{k=-\infty}^{\infty} P_k(t-t_k-\frac{\tau_k}{2}),$$

4.55

then the spectrum is the expected value of Fourier Transform with respect to Poisson arrival time t_k and Rayleigh distributed clip duration τ_k:

$$S(f) = \lim_{T\to\infty}\frac{E_{\tau_k,t_k}|\Im\{r_T\}|^2}{2T}.$$

4.56

Analysis of out-of-band power leakage using the spectrum of clipped signal amounts to integrating Equation 4.56 over the desired frequency band interval. As shown in Figure 4.3 the power amplifier is followed by a filter to control out-of-band leakage. Using the argument presented in the previous section, we can decompose the clipped signal to two components, the original signal and an non-correlated clipped portion as shown in Equation 4.15. Let $H(f)$ denote transfer function of the *linear* filter representing cascade of transmit filter, channel, and receive filter. Then, the spectrum of received signal is

$$S_R(f) = S(f)|H(f)|^2.$$
4.57

Using the same argument as the previous section, conditional bit error rate is a more appropriate measure of clipping distortion. To analyze the effect of clipping on the performance of a receiver using an adjacent channel, assume another similar OFDM system occupies the adjacent channel with a guard interval of Δf_g. For example in wireless LAN OFDM the system channel bandwidth is 100 MHz as shown in Figure 4.10.

The adjacent channel interference due to clipping is

$$ACI = \int \int_{\Delta\omega} |g_\tau(\omega) H_R(\omega - \Delta\omega)|^2 d\omega \, f(\tau) d\tau.$$
4.58

Using the above expression we can calculate conditional error probability $P(e|clip)$ by following the same procedure as previous section.

Following the same reasoning used in the last section to evaluate error probability, treating the above average leakage power to determine the effect on an adjacent channel may be very optimistic at high clipping level. As before, Equation 4.25 describes the effect of a clip on a sub-channel k, only we are now interested in an adjacent system so that $k > N$. That F_k must be multiplied by H_k, the value of the composite filtering function at that sub-carrier frequency.

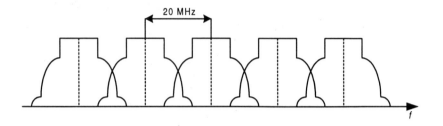

Figure 4.10. Wireless LAN adjacent channels.

F_k again is a probabilistic quantity dependent on τ, which in turn is a function of l. Then numerical techniques must be used to calculate the probability that this quantity exceeds the value which will produce an error for the particular constellation used.

References

1. Mestdagh, D.J.G., Spruyt P.M.P. "A Method to Reduce the Probability of Clipping in DMT-Based Transceivers." *IEEE Trans. Commun.*; Oct 1996; 44: 1234-1238.

2. Shepard, S., Orriss J., Barton S. "Asymptotic Limits in Peak Envelope Reduction by Redundancy Coding in OFDM Modulation." *IEEE Trans. Commun.*; Jan 1998; 46: 5-10.

3. Feig, E., Nadas A. "The Performance of Fourier Transform Division Multiplexing Schemes on Peak Limited Channels." *IEEE Globecom;* 1988; 35.4.

4. Rinne, J., Renfors M. "The Behavior of OFDM Signals in an Amplitude Limiting Channel." *IEEE Int. Conf. Commun.* 1994: 381–385.

5. Mestdagh, D. J. G. , Spruyt P.M.P., Biran B. "Analysis of Clipping Effect in DMT-Based ADSL Systems." *IEEE Int. Conf. Commun.* 1994: 293-300.

6. Gross, R., Veeneman D. "SNR and Spectral Properties for a Clipped DMT ADSL Signal." *IEEE Int. Conf. Commun.* 1994: 843-847.

7. VanVleck, J.H., Middletown D. "The Spectrum of Clipped Noise." *Proc. IEEE*; Jan 1966; 54: 2–19.

8. Mazo, J.E. "Asymptotic Distortion Spectrum of Clipped, DC-Biased, Gaussian Noise." *IEEE Trans. Commun.*, Aug 1992; COM-40; 8: 1339-1344.

9. Minkoff, J. "The Role of AM-to-PM Conversion in Memoryless Nonlinear Systems." *IEEE Trans. Commun.*, Feb 1985; COM-33; 2: 139-144.

10. Rice, S.O. "Distribution of the Duration of Fades in Radio Transmission." *Bell Syst. Tech. J.*; May 1958; 37: 581-635.

11. Kac, M., Slepian D. "Large Excursions of Gaussian Processes." *Ann. Math. Stat.*; Dec 1959; 30: 1215-1228.

12. Cramer, H., Leadbetter M.R. *Stationary and Related Stochastic Processes*. New York: Wiley, 1967.

13. Leadbetter, M.R., Lindgren G., Rootzen H. *Extremes and Related Properties of Random Sequences and Processes*. New York: Springer-Verlag, 1983.

14. Bussgang, J.J. "Crosscorrelation Functions of Amplitude Distorted Gaussian Signals." *Res. Lab. Of Electronics M.I.T., (Cambridge, Mass.) Tech. Rep*; Mar 26, 1952; 216:3.

15. Blachman, N.M. "Detectors, Bandpass Non-Linearities, and Their Optimization: Inversion of the Chebyshev Transform." *IEEE Trans. Inform. Theory*, Jul 1971; IT-17: 398-404.

Chapter 5 *Synchronization*

Multi-carrier modems, like any other digital communication modems, require a reliable synchronization scheme. Parallel transmission of N symbols results in a longer symbol duration, consequently there is less sensitivity to the timing offset. In other words, unlike single carrier systems in which a timing jitter can create inter-symbol interference, it does not violate orthogonality of transmitted waveforms in a multi-carrier system. However, frequency offset is detrimental to OFDM systems and has an important role in system design. Phase noise is another critical impairment in wireless OFDM.

In this chapter we will review the effect of timing and frequency jitter in OFDM, and discuss some of the synchronization techniques.

5.1 Timing and Frequency Offset in OFDM

Assuming that interference between OFDM blocks is negligible, a transmitted OFDM signal can be represented as

$$x(t) = \sum_{k=-k_1}^{N+k_2+1} \sum_{n=0}^{N-1} D_k e^{j2\pi\frac{nk}{N}} w(t - \frac{k}{f_s}), \quad for -k_1/f_s < t < (N+k_2)/f_s,$$

5.1

where k_1 and k_2 are pre-and postfix lengths, and $w(t)$ is the time domain window function. The received signal $r(t)$ is filtered and sampled at the rate of multiples of $1/T$.

The sampled signal at the output of the receiver FFT with ideal channel can be represented as

$$y_n = \left[\sum_{k=-\infty}^{\infty} X_c(f + \frac{Nk}{T}) \right] \otimes TW(fT) \Bigg|_{f=\frac{n}{T}},$$

5.2

where $X_c(f)$ is the Fourier Transform of the periodically repeated analog equivalent of the signal generated by transmitter's IFFT. Therefore, $X_c(f)$ is a line spectrum at $\pm k/T$ and $W(f)$ is the Fourier Transform of window function $w(t)$.

Assuming that the sampling time has a relative phase offset of τ and the offset does not change during one OFDM symbol, the sampled received signal for a non-dispersive channel can be simplified to:

$$y_k = \sum_{n=0}^{N-1} D_n e^{j\varphi} e^{j2\pi\frac{n}{N}f_s t} \Bigg|_{t=\frac{k+\tau}{f_s}},$$

5.3

where ϕ represents envelope delay distortion. After the Fourier Transform at the receiver,

$$\tilde{D}_m = \frac{1}{N} \sum_{k=0}^{N-1} \sum_{n=0}^{N-1} D_n e^{j2\pi\frac{n}{N}k} e^{j\left(\varphi+2\pi\frac{n}{N}\tau\right)} e^{-j2\pi\frac{m}{N}k}$$

$$= \sum_{n=0}^{N-1} D_n e^{j\left(\varphi+2\pi\frac{n}{N}\tau\right)} \sum_{k=0}^{N-1} e^{-j2\pi\frac{k}{N}(n-m)}$$

5.4

$$= \begin{cases} 0, & n \neq m \\ D_m e^{j\left(\varphi + 2\pi \frac{n}{N}\tau\right)}, & n = m \end{cases}$$

As shown, the timing phase, or envelope delay distortion, does not violate the orthogonality of the sub-carriers and the effect of the timing phase offset is a phase rotation which linearly changes with sub-carriers' orders. On the other hand, envelope delay results in same amount of rotation for all sub-carriers. In a more general term, it can be shown [1] that with a pulse shaping filter of roll-off α and in the presence of dispersive channel with impulse response $h(t)$, the detected data is attenuated and phase rotated such that

$$\tilde{D}_m = \gamma_m(\tau) D_m, \qquad\qquad 5.5$$

where

$$\gamma_m(\tau) = \begin{cases} H(\frac{m}{NT})e^{j2\pi\frac{m}{NT}\tau} & , & 0 \leq \frac{m}{N} \leq \frac{1-\alpha}{2} \\[2ex] H(\frac{m}{NT})e^{j2\pi\frac{m}{NT}\tau} + H(\frac{m-N}{NT})e^{j2\pi\frac{m-N}{NT}\tau} & , & \frac{1-\alpha}{2} \leq \frac{m}{N} \leq \frac{1+\alpha}{2} \\[2ex] H(\frac{m-N}{NT})e^{j2\pi\frac{m-N}{NT}\tau} & , & \frac{1+\alpha}{2} \leq \frac{m}{N} \leq 1 \end{cases}$$

$$5.6$$

and $H(m/NT)$ is the Fourier Transform of $h(t)$ at frequency m/NT. Later in this chapter, we discuss estimation of τ and φ.

Frequency offset is a critical factor in OFDM system design. It results in inter-carrier interference and violates the orthogonality of sub-carriers. Equation 5.2, is a mathematical expression of frequency domain samples using a line spectrum representation of the analog signal. Clearly, the effect of frequency offset amounts to inter-channel interference which is similar to inter-symbol interference of a single carrier signal due to a timing jitter. In a non-dispersive channel with rectangular pulse shaping, the interference caused by frequency offset could be too constraining. The sampled signal is

$$y_k = \sum_{n=0}^{N-1} D_k e^{j2\pi\left(\frac{n}{N}f_s+\delta f\right)} \Bigg|_{t=\frac{k}{f_s}}$$

$$= \sum_{n=0}^{N-1} D_k e^{j2\pi\left(\frac{n}{N}+\frac{\delta f}{f_s}\right)k} \qquad\qquad 5.7$$

After DFT we have

$$\tilde{D}_m = D_m \left(\frac{e^{j2\pi\Delta f}-1}{e^{j2\pi\frac{\Delta f}{N}}-1}\right) + \sum_{n=0}^{N-1} D_n \sum_{k=0}^{N-1} e^{j2\pi\frac{k}{N}(n-m+\Delta f)} + N_m, \qquad 5.8$$

where

$$\Delta f = \frac{n\delta f}{f_s} \qquad\qquad 5.9$$

is the relative frequency deviation. As Equation 5.8 shows, In addition to attenuation of desied signal, there is an interference between different symbols of several sub-carriers. This effect creates a significant ICI and is one of the important impairments in OFDM systems, as shown in Figure 5.1. The second term represents the interference from other sub-carriers which is dual to ISI in time domain due to timing offset.

In order to avoid ICI caused by frequency offset, the window function

$$w_n = w(t)\Big|_{t=nT} \qquad\qquad 5.10$$

must be such that zero crossings of its Fourier Transform are at multiples of symbol frequency.

$$W_m = W(\omega)\Big|_{\omega=2\pi n f_s} = \delta_m.$$ 5.11

A generalized *sinc* function of the form

$$\frac{\sin \omega n}{\omega n} \times g(n),$$ 5.12

for any differentiable function, $g(t)$ satisfies the condition of Equation 5.11. Another desired property of a *sinc* function is its low rate of change in the vicinity of sampling points in frequency. One common choice is a raised cosine function in time

$$w_{rc}(t) = \begin{cases} T & , \quad 0 \leq |t| \leq \dfrac{1-\beta}{2T} \\[2mm] \dfrac{\sin \pi \dfrac{t}{T}}{\pi \dfrac{t}{T}} \times \dfrac{\cos \beta \pi \dfrac{t}{T}}{1-4\beta^2 \dfrac{t^2}{T^2}} & , \quad \dfrac{1-\beta}{2T} \leq |t| \leq \dfrac{1+\beta}{2T}, \\[2mm] 0 & , \quad elsewhere \end{cases}$$ 5.13

where β is the roll-off factor for time domain pulse shaping. Notice that a higher roll-off factor requires a longer cyclic extension and guard interval, which consumes higher bandwidth.

Sensitivity of OFDM systems to frequency offset is sometimes too constraining. Therefore, in practice, the zero ICI condition is partly relaxed to provide more robustness against frequency offset. In other words, we tolerate a small amount of ISI at the cost of much less ICI. A Gaussian window is one example in which a single parameter adjustment can control the ISI and ICI trade-off.

The effect of frequency offset on ICI is depicted in Figure 5.1. As shown, proper windowing has a significant impact on the sensitivity of the receiver to frequency offset. A reasonable design technique is to target an acceptable frequency offset and choose a proper window.

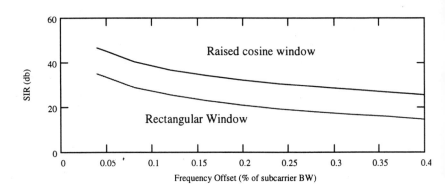

Figure 5.1. ACI caused by frequency offset.

5.2   Synchronization and System Architecture

A synchronization sequence in an OFDM system is shown in Figure 5.2.

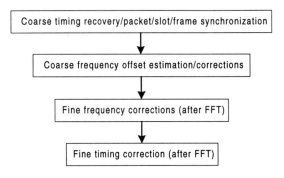

Figure 5.2. Synchronization sequence in OFDM.

Initially, a coarse frame (or packet) synchronization is required to provide timing information regarding the OFDM symbol. Then, a frequency correction prior to FFT is implemented to reduce the effect of inter-channel interference. Consequently, further stages of fine timing and frequency offset are implemented. These may be combined in a joint estimation block. In general, frame timing and frequency correction are more complicated and require higher hardware/software resources. Timing and frequency correction techniques for OFDM can be classified into two categories: data-aided and non-linear techniques. Data aided techniques use a known bit pattern or pilot signal to estimate the timing or frequency offset. Non-linear techniques utilize the cyclostationarity characteristics of the signal to extract the desired harmonic component by using a non-linear operation. Use of data-aided techniques is applicable to many OFDM digital communication systems such as HDTV and Wireless LAN. Pilot signal fields are provided in existing standards.

5.3   Timing and Frame Synchronization

Initially a frame (packet or slot) synchronization circuit is required to detect the frame starting point. This is usually achieved by correlating the incoming signal with a known preamble. The same circuit is sometimes used to adjust for initial gain control. Therefore, the threshold of the detection circuit should be adjusted accordingly. A general block diagram is shown in Figure 5.3. Since timing offset does not violate orthogonality of the symbols it can be compensated after the receiver FFT.

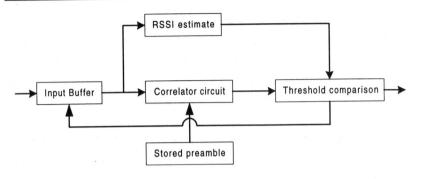

Figure 5.3. Frame synchronization.

As shown in Equation 5.8, the effect of the timing offset is a phase rotation which linearly increases with sub-carrier order. In order to estimate the timing offset, we should solve a regression problem with proper weighting to represent the effect of fading in the channel.

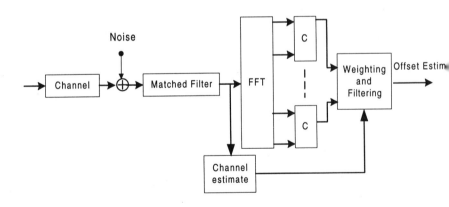

Figure 5.4. Timing offset estimate.

Frequency offset estimation in packet data communications requires fast acquisition. Hence, phase-lock techniques are not applicable for initial coarse correction. For this purpose, a pilot or training sequence should be transmitted. This can be achieved by sending a pilot in one or two bins and estimate the frequency offset by measuring the energy spread in other bins, or by using a training sequence or pilot in every bin and measuring the phase difference by repeating the same pattern. Usually, the second approach is more reliable, as additive noises in different bins are independent and the estimate convergence will be faster. In the following, we discuss both techniques and compare the result analytically and through simulation.

5.4   Frequency Offset Estimation

Frequency offset should be corrected before the receiver FFT. The FFT can be used as a frequency offset detector. Before correction for frequency offset, C/I will not be at the acceptable level for reliable data transmission. Therefore, It is preferrable that entire bins should be used for frequency estimation purposes initially. We assume a known data pattern is used for synchronization. After FFT, at the receiver, we have:

$$D_m = \sum_{k=0}^{N-1} \sum_{n=0}^{N-1} D_n \, e^{j2\pi \frac{k}{N}(n-m+\Delta f)} + N_m. \qquad 5.14$$

If the same block is repeated, the result would be

$$D_m{}^{j2\pi\Delta f} + N_m'. \qquad 5.15$$

Therefore, we can obtain a proper estimate of Δf by averaging the product of every bin from one symbol and the same bin from the previous symbol, and then averaging the result.

$$\frac{1}{N}\sum_{m=0}^{N-1}\left(\frac{1}{l}\sum_{i=0}^{l-1}y_{mi}\,\overline{y}'_{m(i+1)}\right).\qquad\qquad 5.16$$

This process is shown in Figure 5.5:

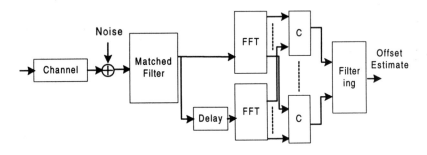

Figure 5.5. Frequency offset estimation.

The number of repetitions depends on the signal-to-noise ratio and the required accuracy of frequency estimation. Double averaging accelerates the convergence of the estimate significantly. This scheme works in low signal-to-noise ratios as well. However, this technique is limited to frequency offset of less than $\Delta f < 1/2T$. The initial frequency offset could be higher than that limit, and requires an initial coarse acquisition. A typical coarse frequency detection in single and multi-carrier systems is the Maximum Likelihood estimation technique.

$$f_{est}=\max_{f}\sum D_{n+l+1}D^{*}_{n+l}.\qquad\qquad 5.17$$

A digital implementation of the technique is shown in the Figure 5.6 where $C(f_i)$ denotes a matched filter tuned to transmit filter impulse response shifted by frequency offset of f_i.

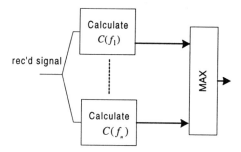

Figure 5.6. A digital implementation of coarse frequency offset estimation using maximum likelihood criteria.

5.5 Phase Noise

Effects of Phase Noise

Intermediate stages of modulation, such as between RF and IF, can introduce degradation due not only to frequency error as described previously, but also to phase noise in the local oscillators involved in such modulation. Phase noise is the zero mean random process of the deviation of the oscillator's phase from a purely periodic function at the average frequency. The oscillator output may be considered to be an ideal sine wave phase modulated by the random process. The power spectral density is generally normalized to the power of the sine wave. Higher quality (and more costly) oscillators will typically have lower phase noise than cheaper ones. Therefore, one may expect greater phase noise at the mobile set in a wireless system, as opposed to the base station, and similarly in the customer terminal as opposed to the central office in a subscriber line transmission system.

An approximation of a typical phase noise power spectral density (PSD) is shown in Figure 5.7:

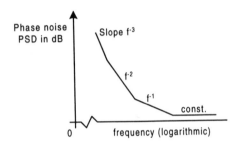

Figure 5.7. A typical oscillator phase noise power spectrum.

The portion with its slope steeper than f^{-2} exists only at a very low frequency. Actually, the PSD may be mathematically inaccurate in that the phase process in this region is typically non-stationary. The f^{-2} slope is equivalent to white frequency noise. The f^{-1} slope in PSD is difficult to model, but agrees with measurements and is commonly referred to as "flicker noise." Finally, the white phase noise at a higher deviation frequency is usually quite low.

The above describes a free-running oscillator. In practice, any oscillator used for intermediate modulation should be phase-locked to eliminate frequency error, for the reasons previously discussed. When a phase-locked loop is employed, the phase noise spectrum is modified. Figure 5.8 shows a classic phase-locked loop and its mathematical model.

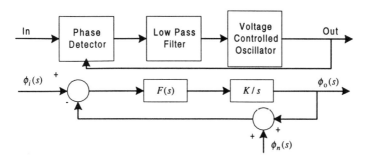

Figure 5.8. Classical phase-locked loop.

The variables Φ are phase deviations from nominal. Φ_i, Φ_o, and Φ_n are the input, output, and internal oscillator phase deviations respectively. The low-pass filter $F(s)$ must be chosen such that the feedback loop is stable. The voltage-controlled oscillator (VCO) has an output frequency deviation from its nominal value that is proportional to the control voltage. Since phase is the integral of frequency, the transfer function of the VCO is K/s as shown. The equation of the loop is

$$\Phi_o = \frac{KF(s)}{s + KF(s)}\Phi_i + \frac{s}{s + KF(s)}\Phi_n.$$ 5.18

Because the open loop transfer function has a pole at zero frequency, the frequency of the VCO will lock exactly to the average input frequency, provided the initial frequency difference is within the lock-in range of the loop. Substituting the frequency variable $j\omega$ for the LaPlace variable s, the power transfer function is

$$|\Phi_o|^2 = \frac{K^2|F(\omega)|^2}{|j\omega + KF(\omega)|^2}|\Phi_i|^2 + \frac{\omega^2}{|j\omega + KF(\omega)|^2}|\Phi_n|^2.$$ 5.19

The phase-locked loop thus acts as a low-pass filter of its input phase noise, and a high-pass filter of the internal VCO noise. Note that this is true even without a low-pass filter [$F(s) = 1$].

More modern phase-locked loops are implemented digitally [1], except at very high frequencies. An example of a digital phase-locked loop is shown in Figure 5.9.

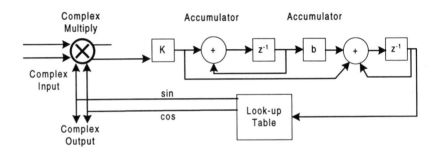

Figure 5.9. A second order digital phase-locked loop.

In this implementation, the input and output are in complex form. The complex multiplier eliminates the need to remove double frequency components, an easy task in analog implementation, but not in digital. Only the imaginary output of the multiplier is used. Instead of a VCO, the same function is performed by an accumulator and phase look-up table. The loop is implemented as a sampled data system. The sampling and computation rate $1/\tau$ need not be particularly accurate or related to the frequency to be locked. As long as τ is much smaller than any of the loop time constants, the operation is equivalent to an analog loop. The particular loop shown is a second order one, where the additional accumulator in the digital filter produces a second pole at zero frequency.

When a phase-locked loop is employed to lock the local oscillator used for intermediate modulation, the resultant phase noise spectrum will be a high-pass filtered version of Figure 5.7, such as the simplified spectrum shown in Figure 5.10 below.

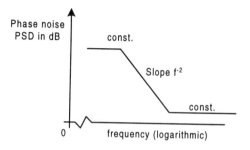

Figure 5.10. Typical spectrum of a phase-locked oscillator.

If the constant high frequency portion can be neglected, then the above (two-sided) PSD is a Lorenzian one,

$$|\Phi_o(f)|^2 = \frac{1}{\pi} \frac{K\beta}{f^2 + \beta^2}.$$ 5.20

The 3 dB bandwidth is β, and the total power is K. The corresponding phase noise is a Weiner process, in which all disjoint phase increments are independent. This noise model is frequently used in analysis of phase noise impairment.

We are now ready to examine the effects of phase noise on an OFDM system [3-6]. Following the same analysis that was used to examine frequency offset, and assuming there are no other impairments, the output of the receiver demodulator is

$$\tilde{D}_m = \frac{1}{N} \sum_{n=0}^{N-1} D_n \sum_{k=0}^{N-1} e^{j\phi(k)} e^{-j2\pi\frac{k}{N}(n-m)}, \ m = 0\cdots(N-1) \qquad 5.21$$

where $\phi(k)$ is the sample of the phase noise at the receiver input at the kth sample time.

If the phase noise is small, as is almost always the case,

$$e^{j\phi(k)} \approx 1 + j\varphi(k), \qquad 5.22$$

so,

$$\tilde{D}_m \approx D_m - \frac{jD_m}{N} \sum_{k=0}^{N-1} \varphi(k) - \frac{j}{N} \sum_{n=0,n\neq m}^{N-1} D_n \sum_{k=0}^{N-1} \varphi(k) e^{-j2\pi\frac{k}{N}(n-m)}. \qquad 5.23$$

The first term in Equation 5.23 is the desired output, the second is a random phase rotation of the sub-carrier, and the third term is inter-carrier interference due to a loss of orthogonality among the sub-carriers.

Let us first examine the second term. When the bandwidth of the phase noise is less than the OFDM symbol rate $1/T$, then

$$\sum_{k=0}^{N-1} \varphi(k) = \frac{N}{T} \int_T \varphi(t)dt, \qquad 5.24$$

or in the frequency domain,

$$\sum_{k=0}^{N-1} \varphi(k) = N \int_{-\infty}^{\infty} \Phi(f) \operatorname{sinc}(fT) df. \qquad 5.25$$

This phase rotation is the same for all sub-carriers. It can readily be eliminated, for example by measuring the phase variation of a pilot sub-carrier and subtracting that rotation from all sub-carriers. Another technique to solve this, and other problems, is the use of differential phase modulation among the sub-carriers. In this technique, one sub-carrier carries a fixed phase and the information on other sub-carriers is carried by phase differences. This is applicable only to constellations of constant amplitude. Furthermore, there is a noise performance penalty when compared with coherent demodulation.

If the above correction is performed, we are left with the inter-carrier interference. Again, assuming the phase noise bandlimited to less than $1/T$, that interference is

$$I = \frac{j}{T} \sum_{n=0, n \neq m}^{N-1} D_n \int_T \varphi(t) e^{j2\pi(n-m)\frac{t}{T}} dt =$$

$$j \sum_{n=0, n \neq m}^{N-1} D_n \int_{-\infty}^{\infty} \Phi(f) \, \text{sinc}(n - m + fT) df.$$

$$5.26$$

If the data symbols on the different sub-carriers are independent, then the interference may be treated as a Gaussian noise of power

$$I^2 = \sum_{n=0, n \neq m}^{N-1} \sigma_n^2 \int_{-\infty}^{\infty} |\Phi(f)|^2 \, \text{sinc}^2(n - m + fT) df, \qquad 5.27$$

where σ_n^2 is the average power of the nth sub-carrier. For any phase noise PSD that is decreasing with frequency, near sub-carriers will contribute greater to this interference than sub-carriers further removed from the sub-carrier of interest. For this reason, sub-carriers near the center of the frequency band will be subject to more interference than sub-carriers at the band edge by up to a factor of two. More importantly, the interference increases as the spacing between sub-carriers decreases.

The noise-to-signal ratio for a sub-carrier near the middle of the band, assuming equal power on all sub-carriers, is

$$\frac{I^2}{\sigma_m^2} = 2 \sum_{n=1}^{N/2-1} \int_{-\infty}^{\infty} |\Phi(f)|^2 \, \text{sinc}^2(p + fT) df. \qquad 5.28$$

For the Lorenzian noise spectrum given previously, this quantity is plotted as a function of βT in Figure 5.11, normalized by the ratio of phase noise power to power in the sub-carrier. Here, all sub-carrier powers were assumed to be equal. Note that βT is the ratio of the 3 dB bandwidth of the phase noise to the OFDM symbol rate. To find the un-normalized noise-to-signal ratio, the plotted quantity should be multiplied by the phase noise to sub-carrier power ratio.

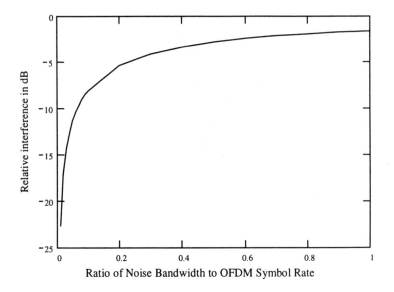

Figure 5.11. Inter-carrier interference noise.

It is seen that as long as the OFDM symbol rate is greater than approximately twice the phase noise bandwidth, the performance degrades as that symbol rate, or, equivalently, the sub-carrier spacing, decreases. Therefore, under these conditions, the use of a large number of sub-carriers to carry a given bit rate is detrimental. However when the OFDM symbol rate increases beyond approximately twice the phase noise bandwidth, there is little further degradation., assuming that the line rate is still much greater than the noise bandwidth.

Another form of phase noise that can exist even in OFDM systems without intermediate modulations, such as in wireline applications, results from jitter in the sampling time at the receiver input. In Reference [7], it is shown that when the timing jitter is constant over an OFDM symbol interval, no inter-carrier interference results. In this case, each sub-carrier constellation is subjected to rotation proportional to the timing offset and to the frequency of that sub-carrier.

References

1. Pollet, T., Moeneclay M. "Synchronizability of OFDM Signals." *IEEE Proc. of Global Telecom Conf. (GLOBECOM '95)*; Nov 1995; 3: 2054-2058.

2. Lindsay, W.C., Chie C.M. "A Survey of Digital Phase-Locked Loops." *Proc. IEEE*; Apr 1981; 9; 4: 410-431.

3. Pollet, T., VanBladel M., Moeneclay M. "BER Sensitivity of OFDM Systems to Carrier Frequency Offset and Wiener Phase Noise." *IEEE Trans. Commun.*; Feb/Mar/Apr 1995; 43; 2/3/4: 191-193.

4. Robertson, P., Kaiser S. "Analysis of the Effects of Phase Noise in OFDM Systems." *IEEE Int. Conf. Commun.*; 1995: 1652-1657.

5. Armada, G., Calvo M. "Phase Noise and Sub-Carrier Spacing Effects on the Performance of an OFDM Communication System." *IEEE Commun. Letters*; Jan 1998; 2; 1: 11-13.

6. Steendam, H., Moeneclay M., Sari H. "The Effects of Carrier Phase Jitter on the Performance of Orthogonal Frequency Division Multiple Access Systems." *IEEE Trans. Commun.*; Apr 1998; 46; 4: 456-459.

7. Zogakis, T.N., Cioffi J.M. "The Effect of Timing Jitter on the Performance of a Discrete Multitone System." *IEEE Trans. Commun.*; Jul 1996; 44; 7: 799-808.

Chapter 6 *Equalization*

6.1 Introduction

One of the advantages of a multi-carrier system is its robustness against inter-symbol interference. The longer duration of OFDM symbols provides higher immunity against delay spread and ISI. As long as channel dispersion is not longer than the OFDM symbol guard interval, system performance does not degrade due to ISI and use of time domain equalization is not usually mandated.

However, in case of higher data rates and channels with extensive time dispersion an equalizer is unavoidable. Albeit, the structure of the equalizer is different from that of single carrier systems. The purpose of equalization is not complete removal but restriction of inter-symbol interference to a tolerable extent.

Frequency domain equalization, in the absence of inter-channel interference (ICI) is used to compensate for channel complex gain at each sub-carrier frequency. Using time-frequency duality, complex gain compensation after FFT is equivalent to a convolution of a FIR filter in time domain (residual equalization).

In this chapter, we review main differences of equalization in single and multi-carrier and follow up with analytical and practical aspects of time domain equalization. Then, frequency domain equalization and echo cancellation are discussed in sections 3 and 4. Using duality prniciple, we present a unified approach to time and frequency domain equalization.

6.2   Time Domain Equalization

A brief review of decision feedback equalizer structure and performance should help the reader to readily understand the analysis and design of equalizers for multi-carrier.

Decision Feedback Equalizers

In a decision feedback equalizer, the forward filter whitens the noise and produces a response with post-cursor ISI only, and the feedback filter then cancels that post-cursor ISI. The optimum forward filter for a zero-forcing DFE can be considered as cascade of a matched filter followed by a linear equalizer and a causal whitening filter whose transfer function can be found by spectral factorization of the channel power spectrum. The forward filter, therefore, first minimizes ISI, **then** noise power at the input of slicer, reintroducing causal ISI. Notice that the separation and ordering has only analytic significance [4]. In practice, all forward filters are usually lumped together in one adaptive filter.

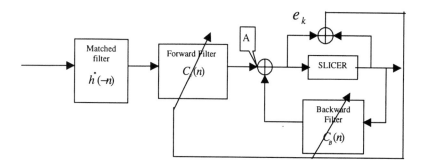

Figure 6.1. General decision feedback equalizer.

In a MSE decision feedback equalizer, the optimum forward filter minimizes the mean-square residual ISI <u>and</u> noise power. In both MSE-DFE and ZF-DFE we intend to minimize the error term at the input of the slicer, which in the absence of error propagation includes residual ISI and noise. Let $h(n)$ represent the channel impulse response, hence $h^*(-n)$ is the impulse response of the matched filter. Forward filter $c_F(n)$ performs ISI suppression and noise whitening. Feedback filter $c_B(n)$ eliminates the interference of previous symbols, hence $1 + c_B(n)$ has to be a monic and causal filter. Therefore, its output should have post-cursor inter-symbol interference only, namely

$$h(n) * h^*(-n) * c_F(n) = 1 + c_B(n). \qquad 6.1$$

On the other hand the variance of the noise, at the output of the matched filter is

$$\sigma^2 \int_{-\pi/2}^{\pi/2} S_h(e^{j\omega}) d\omega \qquad 6.2$$

where S_h is power spectrum of the channel impulse response. In order to minimize noise power at the input of the slicer, $c_F(n)$ should whiten the noise. Rewriting Equation 6.1 in the frequency domain, the forward filter transfer function is

$$C_F(z) = \frac{1 + C_B(z)}{H(z)H^{*-1}(z^{-1})}.$$ 6.3

The forward filter may apply further linear transformation to the input signal to meet some optimality criteria such as minimizing square error or peak distortion. To whiten a stochastic process with spectrum of $\sigma^2 S_h(z)$ it should satisfy the following equation:

$$\sigma^2 S_h(z) \frac{1 + C_B(z)}{H(z)H^{*-1}(z^{-1})} \frac{1 + C_B^*(z^{*-1})}{H^*(z^{*-1})H(z)} = \sigma^2_{DFE}.$$ 6.4

Since

$$S_h(z) = H(z)H^{*-1}(z^{-1})$$ 6.5

from Equation 6.4 we have:

$$S_h(z) = (1 + C_B(z))(1 + C_B^*(z^{*-1})).$$ 6.6

The same procedure [4] can be followed for the MSE equalizer to obtain

$$S'_h(z) = (1 + C_B(z))(1 + C_B^*(z^{*-1})).$$ 6.7

In the following, we discuss some important issues about the DFE with finite tap FIR feedback and forward filters as a preface for future discussions.

The input to the feedback filter is noise free, therefore the feedback filter eliminates the post-cursor ISI both in ZF-DFE and MSE-DFE. The error term used for training is memoryless and defined as:

$$e_k = x_k - \hat{x}_k.$$ 6.8

where x_k is transmitted symbol, which is the same as the output of the slicer when there is no error and $\hat{x}_k$ is the output of the equalizer. Minimizing $|e_k|^2$ requires that input signal be orthogonal to the error term. Since the input to the feedback filter is noise free, we design the forward filter to minimize the error power which leads to the following equations:

$$E(\varepsilon_k y_{k-i}^*) = E(x_k - \sum_j c_{F_j} y_{k-j} - \sum_{j'} c_{B_{j'}} x_{k-j'}) y_{k-i}^* = 0 \quad for \quad -\infty < i < \infty,$$

$$6.9$$

where:

$$y_k = \sum_{n=0} p_n x_{k-n} + \eta_k.$$

$$6.10$$

and

$$p_i = \sum_{n=0}^{l-i} h_n h_{n+i}^* \qquad i = 0, \cdots, l$$

$$p_{-i} = p_i^*.$$

where l is channel spread in symbols.

Therefore,

$$\sum c_{F_j} E\{\sum_n p_n x_{k-n-j} \sum_{n'} p_{n'}^* x_{k-n'-i}^*\} + \sum_j c_{F_j} E\{\eta_{k-j} \eta_k^*\}$$

$$= \sum_{n=0}^{N} p_n E(x_k x_{k-n-i})$$

$$\Rightarrow \sum_j c_{F_j} \sum_n p_n p_{n+j-i}^* + N_0 \sum_j c_{F_j} p_j = p_{-i}^*.$$

$$6.11$$

η_k is a colored Gaussian noise with the following auto-correlation function

$$E(\eta_i \eta_j^*) = \begin{cases} N_0 p_{i-j}, & |i-j| \le l \\ 0, & otherwise \end{cases},$$ 6.12

The solution to equation 6.9 is

$$C_F = \Gamma^{-1}\theta,$$ 6.13

where

$$\Gamma_{ij} = E\{y_{k+i}y_{k+j}^*\}, \qquad 0 \le i, j < N.$$ 6.14

N is the number of forward taps and

$$\theta_i = E\{x_k y_{k+i}\}, \qquad 0 \le i < N.$$ 6.15

Coefficients of the backward filter can be found by

$$C_B = P^T C_F,$$ 6.16

where

$$C_F = \begin{pmatrix} c_{F_0} \\ c_{F_{-1}} \\ \\ c_{F_{-N}} \end{pmatrix} \quad C_B = \begin{pmatrix} c_{B_1} \\ c_{B_2} \\ \\ c_{B_M} \end{pmatrix} \quad P = \begin{pmatrix} p_0 & p_1 & \cdots & p_M \\ p_1 & p_2 & & p_{M+1} \\ & & & \\ p_N & & & p_{N+M} \end{pmatrix}.$$ 6.17

The forward filter c_F is anti-causal which means that the main tap is assumed to be the closest one to the slicer. The forward taps should remove the entire post-cursor. Practically, the matched filter is usually absorbed in the forward equalizer[4] then h replaces p is Equations 6.12 and 6.16. Notice that the feedback filter is designed to minimize the post-cursor ISI at the input of the slicer. In practice, c_B and c_F are adjusted adaptively.

---

[4] We can interpret this as replacing the matched filter with a whitening matched filter

6.3   Equalization in DMT

Multi-carrier systems are robust against inter-symbol interference as long as orthogonality of adjacent symbols is preserved in the frequency domain. In other words, the symbol duration of the OFDM signal is extended to beyond *T*. However, if channel impulse response spread is more than the prefix extension, inter-symbol interference can degrade the performance.

There are several fundamental differences between time domain equalization requirements of OFDM channels and classic decision feedback equalizers:

a)   An equalizer of an OFDM system does not need to cancel the ISI entirely but to limit its length. Unlike a single carrier system in which equalization minimizes ISI, we only need to reduce it to a time span less than the length of guard interval.

b)   The channel impulse response is modelled as an ARMA system. Usually, the channel impulse response is too long. Therefore, a model of the form

$$\frac{A(z)}{1 + B(z)} \qquad 6.18$$

is appropriate.

c)   The equalizer does not work as a DFE in data mode. Unless channel characteristics change rapidly, equalizer taps are adjusted during the training sequence and the same tap values are preserved during data mode.

d) The feedback filter is not in the loop and need not to be monic or causal. In single carrier systems, the slicer output is fed to the feedback filter to remove the post-cursor effect.   However, truncating the impulse response does not require a monic feedback filter. A typical structure is shown in Figure 6.2.

e) The error is not memoryless. The error term used in typical DFE is derived from the scalar input and output of the slicer. In a DMT equalizer the error is derived from the output of an impulse shortening filter and reference filter.

f) The location of the impulse response window has a big impact on the performance of equalizer.

The overall structure of the equalizer in training mode is shown in the following diagram:

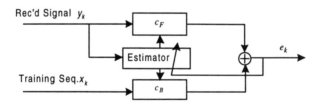

Figure 6.2. Equalizer configuration in training mode.

In steady state mode, the equalizer will include only c_F. The error term is

$$e_k = c_F * Y_k - c_B * X_k.$$
6.19

where X_k and Y_k are vectors of training and received samples respectively. In general, we can use adaptive techniques to find near optimum equalizer coefficients. A large number of taps prevents us from using RLS type of algorithms. On the other hand, a wide spread of eigenvalues of the input signal covariance matrix can slow down convergence of the equalizer. A different technique for equalizer tap adjustment is based on linear prediction theory as explained below.

During acquisition, the frame is repeated and can be used for equalizer training. In this approach, we use second order statistics of the channel to produce a one shot estimate of equalizer coefficients. In order to use the predictor format of linear regression techniques, without loss of generality, assume c_F is monic. So, the predictor format is:

$$\hat{y}(n|n-1) = \varphi^T(n)C \qquad 6.20$$

where

$$\varphi(n) = [-y(n-1), -y(n-2), \cdots, -y(t-N), x(n-1), x(n-2), \cdots, x(n-M)]^T \qquad 6.21$$

and,

$$C = [C_F \ C_B]^T. \qquad 6.22$$

The least square solution of linear regression is:

$$\hat{C} = \Gamma^{-1}\theta \qquad 6.23$$

with:

$$\Gamma = \begin{pmatrix} A_{N \times N} & B_{N \times M} \\ B^*_{M \times N} & I_{M \times M} \end{pmatrix}, \qquad 6.24$$

where

$$A_{ij} = R_{yy}(i-j) = \frac{1}{N} \sum_{n=1}^{N} y(n-i)y(n-j), \qquad 6.25$$

and

$$B_{ij} = R_{xy}(i-j) = \frac{1}{N} \sum_{n=1}^{N} x(n-i)y(n-j), \qquad 6.26$$

and

$$\theta = [D_N \quad E_M]^T \qquad\qquad 6.27$$

with

$$D_i = R_{yy}[i], \quad E_i = R_{yx}[i], \qquad\qquad 6.28$$

where R_{yy} and R_{xy} are auto-correlation and cross correlation functions. We can use channel impulse response to calculate coefficients directly

$$A_{ij} = \sum_n h_n h^*_{n+j-i} \qquad\qquad 6.29$$

$$B_{ij} = h^*_{-i+j}. \qquad\qquad 6.30$$

Estimation of channel impulse response can help in choosing optimum window location. As explained before, c_B is not necessarily causal, therefore dely parameter plays an important role in the performance of the equalizer. Other important parameters are the lengths of the two filters c_F and c_B.

Delay Parameter

Unlike classic decision feedback equalizers, an ARMA equalizer can have a feedback filter with non-causal transfer function i.e. the general error term is:

$$c_F * Y_k - c_B * X_{k-d}. \qquad\qquad 6.31$$

Delay unit d has significant effect on overall performance. The number of taps M of filter c_B is approximately the same as the prefix length and number of taps for the forward filter $N \leq M$. The delay unit should be chosen properly to capture the most significant window of the impulse response. Brute force trial and error is one option. A more intelligent technique can use the estimate of the impulse response to pick a proper starting point where the energy of the impulse response is above a threshold. The threshold should be the ratio of tap power to overall window power. Since the length of the window function and forward equalizer should be as short as possible, proper choice of starting point is of significant importance. In other words, in an ARMA model the number of zeroes is not assigned properly.

AR Approximation of ARMA Model

Direct solution of Equation 6.23 requires a non-Toeplitz matrix inversion and is numerically sensitive and intensive. One technique to alleviate the numerical difficulty of least square solution is by using techniques for AR approximation of the channel response ARMA model, which results in a numerically attractive Toeplitz correlation matrix. A description of embedding techniques is presented in the Appendix. In the following, the procedure for this analysis is briefly described.

By using embedding techniques, as explained in the Appendix, the ARMA model of Equation 6.18 is approximated by a two-channel AR model.

If the covariance matrix is available the problem can be reduced to solution of a normal (Yule-Walker) equation. The multi-channel version of the Levinson algorithm is a computationally efficient solution for the two-channel AR model at hand. As a reference, the recursive technique is presented here:

Initialization:

$$C_i^f[1] = -R^{-1}[0]R[1]$$
$$C_1^b[1] = -R^{-1}[0]R[-1]$$

6.32

where

$$C_i^f[j] = \begin{bmatrix} -a_i^j & b_i^j \\ 0 & 0 \end{bmatrix}$$

is the j-th forward prediction coefficient of two-channel model at i-th iteration and $C_i^b[j]$ is the j-th backward prediction coefficient at i-th iteration and

$$R[i] = \begin{bmatrix} R_{yy}[i] & R_{yx}[i] \\ R_{xy}[i] & R_{xx}[i] \end{bmatrix}$$

If the input signal x is white then

$$R[i] = \begin{bmatrix} R_{yy}[i] & R_{yx}[i] \\ 0 & 0 \end{bmatrix}$$

with

$$R[0] = \begin{bmatrix} R_{yy}[0] & R_{yx}[0] \\ R_{xy}[0] & I \end{bmatrix}$$

Recursion:

$$C_i^f[i] = K_i^f$$
$$C_k^f[i] = C_{k-1}^f + K_k^f[i]C_{k-1}^b[k-i]$$
$$C_i^b[i] = K_i^b$$
$$C_k^b[i] = C_{k-1}^b + K_k^b[i]C_{k-1}^f[k-i]$$

6.33

where the reflection coefficients are,

$$K_k^f = -\Delta_k^{f^T} \Sigma_{k-1}^{b^{-T}}$$
$$K_k^b = -\Delta_k^{f^T} \Sigma_{k-1}^{b^{-T}}$$
$$\Delta_k^f = \sum_{i=0}^{k-1} R[k-i] C_{k-1}^{f^T}[i]. \qquad 6.34$$

Figure 6.3 shows the structure of an equalizer using a lattice filter. This structure represents the case of equal number of poles and zeros. The extension to unequal number of poles and zeroes is straightforward and explained in [5].

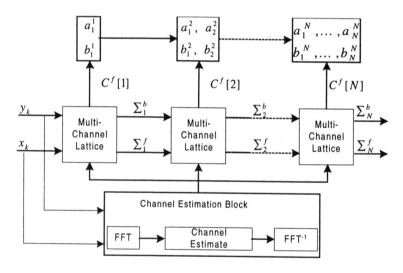

Figure 6.3. Equalization using order update.

The above approach is convenient for batch or frame based processing where the entire frame is buffered and processed for equalization. Instead of inverting the large matrix of Equation 6.13 for example, this structure can be used in the training mode of an ADSL receiver. This technique can be implemented iteratively in time and order when the covariance is not known in advance. The time/order updating algorithm increases the order while updating the coefficients of the equalizer. In some cases, we can preserve the order and continue updating the parameters using adaptive techniques. This technique is useful when the delay profile of the channel does not change significantly but some slow adaptation improves the performance.

6.4   Frequency Domain Equalization

Once it is assured that orthogonality of the sub-carriers is maintained, possibly through the use of the cyclic prefix and time domain equalization as previously discussed, then the final frequency domain equalization of an OFDM signal is an extremely simple process. This is certainly one of the key advantages of OFDM.

After demodulation, the sub-carriers will be subjected to different losses and phase shifts, but there will be no interaction among them. Frequency domain equalization therefore consists solely of separate adjustments of sub-carrier gain and phase, or equivalently of adjusting the individual decision regions. For the case where the constellations consist of equal amplitude points, as in PSK, this equalization becomes even simpler in that only phase need be corrected for each sub-carrier, because amplitude has no effect on decisions.

A simplified picture of the place of frequency domain equalization is shown in Figure 6.4, where the equalizer consists of the set of complex multipliers, {A}, one for each sub-carrier.

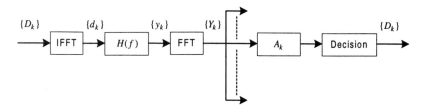

Figure 6.4. An OFDM system with frequency domain equalization.

Here the linear channel transfer function $H(f)$ includes the channel, the transmit and receive filters, and any time domain equalization if present. $H(f)$ is assumed bandlimited to less than N/T for a complex channel. Cyclic extension is not shown although it is almost certain to be present. The following analysis assumes that both amplitude and phase need be corrected, and that equalization consists of multiplying each demodulated component by a quantity such that a fixed set of decision regions may be used.

The signal presented to the demodulator is

$$y(t) = \sum_{n=0}^{N-1} d_n h(t - n\frac{T}{N}).$$ 6.35

The kth output of the demodulator is then

$$Y_k(f) = D_k H_k \; , \quad k = 0,1,...N-1,$$ 6.36

where the H_k are samples of $H(f)$

$$H_k = H(\frac{k}{T}).$$ 6.37

Thus each output is equal to its associated input data symbol multiplied by a complex quantity which differs among the outputs, but are uncoupled. Equalization at its simplest then consists of setting the multipliers to $1/H_k$ for each non-zero channel.

The above approach is optimum in every sense under high signal-to-noise conditions. It also produces minimum probability of error at any noise level, and is an unbiased estimator of the input data D_k. However if the criterion to be optimized is minimum mean-square error particularly, then the optimum multipliers are modified to

$$A_k = \frac{1}{H_k} \frac{1}{1 + \dfrac{\sigma_k^2}{|D_k H_k|^2}},$$ 6.38

where σ_k^2 is the noise power in the demodulated sub-channel. However this value produces a biased estimator and does not minimize error probability.

As a practical implementation issue, for variable amplitude constellations, it is frequently desirable to have a fixed grid of decision regions. The A_k s can then be scaled in amplitude such that the separation of constellation points is constant.

Since a shift in timing phase τ is equivalent to a phase shift

$$H_k = e^{j2\pi \frac{k}{T}\tau}$$ 6.39

frequency domain equalization readily corrects for such timing shift.

In principle frequency domain equalization could be employed when orthogonality is lost due to interference among OFDM symbols. In this case, rather than a simple multiplier per sub-channel, a matrix multiplication would be required. This approach is bound to require more computational load than the combination of time and frequency domain equalizers.

During system initialization, any time domain equalizer must be adjusted before frequency domain equalization is performed. Then any periodic test signal with full frequency content, such as a repeated segment of a PN sequence without cyclic prefix, may be used to adapt the frequency domain equalizer. The demodulator outputs should be averaged over many periods to minimize the effects of noise, and the reciprocal of these measured values used to set the multipliers. During steady state data operation adaptation may be performed using a single variable LMS algorithm as shown below.

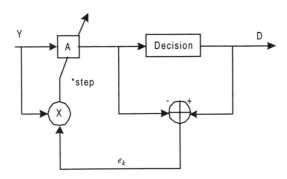

Figure 6.5. LMS adaptation of a frequency domain equalizer multiplier.

This LMS algorithm will adapt to the values given by Equation 6.38 above.

An interesting interpretation of time domain and frequency domain equalizer can be obtained by studying their role in channel distortion compensation. As discussed before, a typical channel model for DMT and channels with long impulse response is an ARMA model of the form 6.18 Time domain equalization shortens the impulse response to a tolerable level for an OFDM system. Mathematically, it's equivalent to compensating the AR part of the channel impulse response $1/(1+B(z))$. So after successful time domain equalization, the equivalent impulse response of channel is reduced to a FIR filter of short duration $A(z)$. Since it does not violate orthogonality of sub-carriers, we can remove its effect after the FFT by frequency domain equalization. The above process is shown in Figure 6.6.

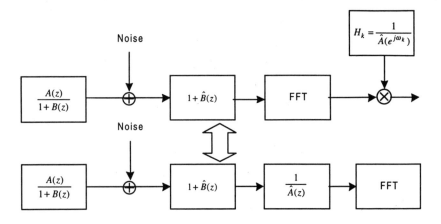

Figure 6.6. Time and frequency domain equalization.

6.5   Echo Cancellation

In wireline applications, it is usually highly desirable to communicate full duplex (simultaneous transmission in both directions) over a single wire-pair. A classic sub-system known as the hybrid provides partial separation, but is not sufficient for high performance digital applications.

Two possible techniques are frequency division and time division duplexing. In the former, different frequency bands are used in each direction, with analog filtering to provide separation. In the latter, alternating time slots are used to transmit alternately in each direction. For symmetric communication, that is equal rates for the two directions, both of these techniques require at least double the bandwidth as unidirectional transmission, which is a very severe performance penalty.

Far preferable is the technique of echo cancellation, where at each end the leakage of transmitted signal into the co-located receiver is cancelled by creating and subtracting a replica of that leakage. The precision required is quite high, as illustrated in the example of a bad case wireline system shown in Figure 6.7.

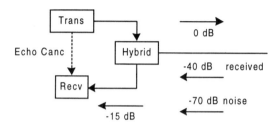

Figure 6.7. An example system illustrating echo cancellation requirement.

In this example, the loss of the line is 40 dB. while the hybrid provides only 15 dB of loss. The leakage of the transmitted signal, or "echo", is therefore 25 dB higher than the desired received signal. One consequence is that the level which the front end of the receiver, in particular the A/D converter, must accommodate is dominated by the echo rather than the received signal. Of more interest here is the high degree of echo cancellation required. Assume, as in the figure, that a signal-to noise ratio of 30 dB is required, and that uncancelled echo can be treated as noise. If we use the reasonable design rule that the uncancelled echo should be at least 10 dB below the line noise, thereby increasing the noise by less than 0.4 dB, the level of the uncancelled echo must be below -80 dB. Because the hybrid provides only 15 dB of reduction, the echo canceller must provide an additional 65 dB of rejection. This requirement is far more severe than that of equalization, so that a much greater arithmetic precision and number of taps is required.

In principle, the same form of echo canceller used in single carrier systems could be used in OFDM. This is shown in Figure 6.8.

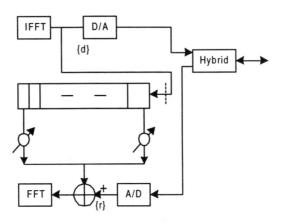

Figure 6.8. A basic echo canceller.

Residual echo is minimized when the tap coefficients of the echo canceller are equal to the corresponding samples of the echo path response $p_k = p(kT)$, where the sampled echo is

$$r_n = r(nT) = \sum_k d_k p_{n-k}.$$
                                                          6.40

Adaptation is similar to that of equalization. Residual error results from echo samples beyond the time span of the canceller, and also from finite precision in signal sampling and in the coefficients. Unlike the case of single carrier equalization, the inputs to the canceller are samples of signals quantized by an A/D converter as opposed to exact data symbols. In one variation [6], adaptation is performed in the frequency domain, followed by an IFFT and time domain cancellation.

In any variation, time domain cancellation requires extensive computation, reducing a prime advantage of OFDM. Another possibility is frequency domain cancellation. If the echo of the transmitted OFDM signal maintains orthogonality, then echo cancellation may be performed very easily after demodulation in the receiver. An advantage of this approach is the use of exact data values as inputs to the canceller rather than quantized signal samples. Furthermore, orthogonality leads to simple adaptation. The canceller consists of one complex multiplication per sub-carrier, as in the frequency domain equalizer.

In order for this technique to be valid, there must be no inter-symbol interference among OFDM signals. Otherwise a matrix multiplication is required rather than a single multiply per sub-carrier [7]. In another approach described in Reference [8], the time domain equalization is modified so as to shorten both the channel response and the echo response to a time span less than the duration of the cyclic prefix. The resulting equalizer and echo canceller can then both be performed quite simply in the frequency domain as shown in Figure 6.9.

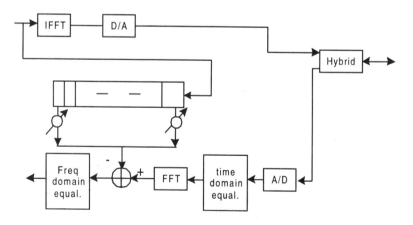

Figure 6.9. Frequency domain echo cancellation.

Techniques have been proposed for echo cancellation in OFDM that combine time and frequency domain components in ways to reduce total required computation [8, 10]. Shown below is the approach of Reference [8] for a symmetric system.

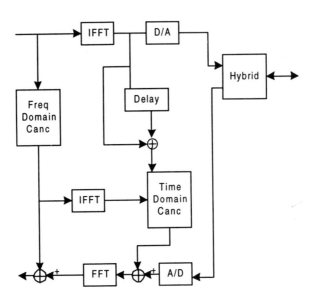

Figure 6.10. Combination echo cancellation for symmetric transmission.

The time domain canceller is a short one if the echo response is not much longer than the duration of the cyclic prefix. Its functions are to undo the effect of cyclic extension, and to cancel the overlap from the previous symbol. The frequency domain canceller can now deal with orthogonal symbols as before.

In asymmetric applications such as ADSL, the penalty for using frequency division duplexing is not as great because the bandwidth used in one direction is relatively small. However for optimum performance, the use of overlapping bands and echo cancellation is employed. Shown in Figure

6.11 and Figure 6.12 are modifications of the previous structure [8] when the same OFDM symbol rate but different numbers of sub-carriers are used in each direction, with overlap of the frequencies of some of the sub-carriers.

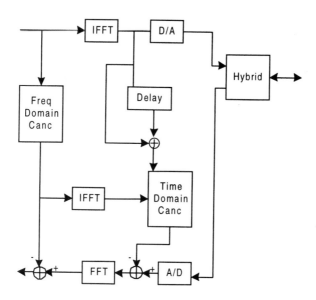

Figure 6.11. Echo cancellation where the transmit rate is lower than receive rate.

Figure 6.11 illustrates a case in which the transmit rate is lower than the receive rate, as in the remote terminal of ADSL. Because the transmit spectrum is smaller than the receive spectrum, it is replicated so that the number of sub-carriers is increased to equal that of the received signal. For the time domain portion of the echo canceller, the transmit signal is interpolated to the higher rate by inserting the appropriate number of zeroes between the time samples.

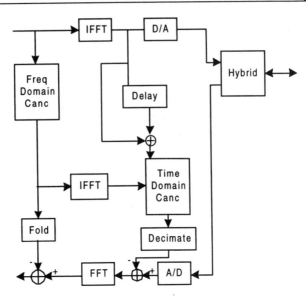

Figure 6.12. Echo cancellation where the receive rate is lower than transmit
rate.

The converse case, in which the transmit rate is higher than the receive
rate, as in the central office terminal of an ADSL link, is shown Figure 6.12.
The size of the input to the frequency domain canceller is reduced by folding
subsets of the OFDM symbol on top of each other and adding them to
produce a number of sub-carriers equal to the number received. The time
domain sampling rate is reduced by computing the short echo at further
spaced points.

Initial adaptation, for any echo cancellation structure, may be performed
at the same time and with the same training signals as the equalization at the
other end. Having the remote transmitter silent improves this initial
adaptation.

6.6   Appendix — Joint Innovation Representation of ARMA Models

Some of the least squares estimation algorithms developed for AR models such as ladder form realization, can be extended to ARMA models by using embedding technique. Namely, an ARMA model can be embedded into a two channel AR model of joint input/output process [11]. The same approach can be used to model an equalizer for an ARMA channel by extending the results of the equalizer design for AR channels. Unlike AR models, the covariance matrix of an ARMA model is not Toeplitz and that results in some numerical computational difficulties. Therefore, by using embedding techniques we can transform an ARMA model to a multi-channel AR model with a block-Toeplitz covariance matrix.

First, we present a geometric interpretation of prediction techniques for ARMA processes which is the basis for understanding Levinson Algorithm.

In linear prediction theory, we predict x_n by $\hat{x}_n^m$, a linear combination of previous m samples $\{x_{n-1}, x_{n-2}, \cdots, x_{n-m}\}$ as a m-th order linear predictor, such that an error criteria like

$$\mathrm{E}\{|x_n - \hat{x}_n^m|^2\} \qquad\qquad 6.41$$

is minimized. The concept of an innovation process is a critical one in prediction and estimation. It represents the amount of new information obtained by observing a new sample x_n which was not available in previous observations:

$$u_n = x_n - x_{n|n-1} \qquad\qquad 6.42$$

Estimation of sample x_n using the previous m samples leads to an mth order forward linear predictor

$$\hat{x}_{n|n-1}^{m} = -\sum_{k=1}^{m} c\ [k]x_{n-k} \qquad\qquad 6.43$$

while using future m samples leads to a backward mth order linear predictor

$$\hat{x}_{n|n+1}^{m} = -\sum_{k=1}^{m} c\ [k]x_{n+k}. \qquad\qquad 6.44$$

Error terms for backward and forward predictors represent an innovation process. These error terms are defined as:

$$e_{f}^{m}[n] = x_{n} - \sum_{k=1}^{m-1} c[k]x_{n-k}$$

$$\qquad\qquad\qquad\qquad\qquad 6.45$$

$$e_{b}^{m}[n] = x_{n-m} - \sum_{k=1}^{m-1} c[k]x_{n-m+k}.$$

A recursive implementation of linear predictors utilizes Gram-Schmidt orthogonalization to transform samples x_{n} into orthogonal random variables [12].

By interpretation of the expression in Equation 6.41 as a metric of the Hilbert space spanned by process x, a useful geometrical representation for linear prediction can be offered as shown in Figure 6.13.

Next we briefly review the Levinson algorithm which is the origin of important concepts such as lattice filters and equalizers. Originally, the Levinson algorithm was proposed as an efficient numerical technique for inverting toeplitz matrices.

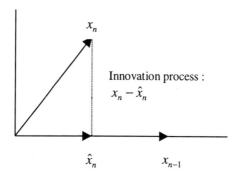

Figure 6.13. Geometric interpretation of linear predictors.

Its order-recursive nature leads to important concepts in estimation and equalization. Minimizing the error criteria 6.41 for the predictor of order m results in the normal equation

$$\Phi_m c_m = \phi_m \qquad\qquad 6.46$$

where Φ_m is Toeplitz matrix of correlations $\Phi_{kl} = \phi(k - l)$ with

$$\phi(n) = E\{x_l x_{l+n}\}. \qquad\qquad 6.47$$

The Levinson algorithm calculates the n coefficients of n-th order linear predictor using n-1 coefficients of $(n$-$1)$th order linear predictor. As a reference we provide the scalar Levinson equations:

Initialization:

$$c_1[1] = \frac{-\phi[1]}{\phi[0]}, \quad e_0 = \phi[0] \qquad\qquad 6.48$$

Recursion:

$$c_k[k] = -\frac{\phi[k] + \sum_{l=1}^{k-1} c_{k-1}[l]\phi[k-l]}{\varepsilon_{k-1}}$$

$$c_k[i] = c_{k-1}[i] + c_k[k]c_{k-1}^*[k-i] \qquad 6.49$$

$$e_k = (1-|c_k[k]|^2)\, e_{k-1}$$

where $c_k[i]$ represents i-th coefficient of he k-th order predictor. In the above equations, we did not explicitly distinguish between backward and forward coefficients because optimal backward coefficients are reversed in time and complex conjugate of optimal forward predictor coefficients.

When the exact order of the predictor is unknown, we can monitor the error term e_k after each iteration. An extensive treatment of this issue is discussed in model order estimation theory.

The above order recursive technique is the basis of lattice representation of prediction filter as shown in Figure 6.14.

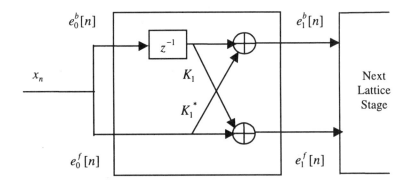

Figure 6.14. Lattice filter representation of predictors.

In general, an ARMA model can be represented as:

$$y_k + b_1 y_{k-1} + \cdots + b_N y_{k-N} = a_0 + a_1 x_k + \cdots + a_N x_{k-N} \qquad 6.50$$

where x_k is the input process and y_k represents the output process. The innovation process of y_k which represents additional information obtained from observation of sample y_k is:

$$e_{k,N} \equiv y_k - \hat{y}_{k|k-1} = [B_N \quad -A_N] \begin{bmatrix} Y_N \\ X_N \end{bmatrix} \qquad 6.51$$

where

$$X^T{}_N = [x_k, \cdots, x_{k-N}]$$
$$Y^T{}_N = [y_k, \cdots, y_{k-N}]$$
$$A^T{}_N = [a_0, \cdots, a_N]$$
$$B^T{}_N = [1, b_1, \cdots, b_N]$$

The problem of linear prediction is: given a set of observed samples of $\{y_{n-1}, y_{n-2}, \cdots, y_{n-m}, x_{n-1}, \cdots, x_{n-m}\}$ an estimate of unobserved sample y_n is required, usually without covariance information. The prediction filter is chosen to minimize the power of prediction error which means designing a filter of the form

$$H_N(z) = \frac{a_0 z^N + a_1 z^{N-1} + \cdots + a_N}{z^N + b_1 z^{N-1} + \cdots + b_N} \qquad 6.52$$

such that

$$y(z) = H_N(z) e_{k,N}(z) \qquad 6.53$$

where $e_{k,N}$ is as close to a white noise process as possible.

One approach to linear prediction of ARMA models is to introduce an innovation process for input x_k, then embed it into a two-channel AR process. The general approach for embedding is discussed extensively in the literature. Here we discuss the embedding technique for the special case where x_k is white. By introducing the augmented process $w_k = [y_k \quad x_k]$, the new prediction error is

$$e_{k,N} = C_N w_k,$$ 6.54

where

$$e_{k,N} = \begin{bmatrix} e_N^x \\ e_N^y \end{bmatrix} = \begin{bmatrix} a_0 x_k \\ x_k \end{bmatrix}$$ 6.55

and

$$C_N = \left[\begin{bmatrix} I & 0 \\ 0 & I \end{bmatrix}, \begin{bmatrix} b_1 & -a_1 \\ 0 & 0 \end{bmatrix}, \dots, \begin{bmatrix} b_N & -a_N \\ 0 & 0 \end{bmatrix} \right].$$ 6.56

Geometric interpretation of the ARMA model using the embedded technique is straightforward and useful. In the ARMA model the space of observations is spanned by augmented process $[x, y]$. Then the joint backward and forward innovations can be interpreted as orthogonal projections on the sub-spaces spanned by future and past observation respectively.

The Levinson recursion algorithm can be applied to the augmented (multi-channel) system as well. However, unlike the scalar case, backward and forward predictor coefficients are not time-reversed and complex conjugate versions of each other.

References

1.   Bingham, J.A.C. "Multicarrier Modulation for Data Transmission: an Idea whose Time has Come." *IEEE Commun. Mag.*: May 1990; 28: 5-14.

2.   Chow, J., Cioffi J., Bingham J. "Equalizer Training Algorithms for Multicarrier Modulation Systems." *IEEE Int. Conf. Commun*; 1993: 761-765.

3. Falconer, D., Magee F. "Adaptive Channel Memory Truncation for Maximum Likelihood Sequence Estimation." *Bell System Tech. Journal*; Nov 1973; 52: 1541-1562.

4. Lee, E.A., Messerschmitt D.G. *Digital Communication.* Boston: Kluwer Academic Publisher, 1994.

5. Al-Dhahir, N., Cioffi J. "Optimum Finite Length Equalization for Multicarrier Transceiver." *IEEE Globecom*; 1994: 1884-1888.

6. Cioffi, J.M., Bingham J.A.C. "A Data Driven Multitone Echo Canceller." *IEEE Trans. Commun.*; Oct 1994; 42; 10: 2853-2867.

7. Yang, J., Roy S., Lewis N.H. "Data-Driven Echo Cancellation for a Multitone Modulation System." *IEEE Trans. Commun.*; May 1996;44; 5: 2134-2144.

8. Melsa, P.J.W., Younce R.C., Rohrs C.E. "Impulse Response Shortening for Discrete Multitone Transceivers." *IEEE Trans. Commun.*; Dec 1996; 44; 12: 1662-1672.

9. Ho, M., Cioffi J.M., Bingham J.A.C. "High-Speed Full-Duplex Echo Cancellation for Multitone Modulation." *IEEE Int. Conf. Commun.*; 1993: 772-76.

10. Ho, M., Cioffi J.M., Bingham J.A.C. "Discrete Multitone Echo Cancellation." *IEEE Trans. Commun.*; Jul 1996; 44; 7: 817-825.

11. Lee, D. L., Friedlander B., Morf M. "Recursive Ladder Algorithms for ARMA Modeling." *IEEE Trans. Automatic Control*, Aug 1982; AC-27: 4.

12. Kay, S.M. *Modern Spectral Estimation.* Upper Saddle River, NJ: Prentice Hall, 1988.

Chapter 7 *Channel Coding*

7.1 Need for Coding

In almost all applications of multi-carrier modulation, satisfactory performance cannot be achieved without the addition of some form of coding. In wireless systems subjected to fading, extremely high signal-to-noise ratios are required to achieve reasonable error probability. In addition, interference from other wireless channels is frequently severe. On wireline systems, large constellation sizes are commonly employed to achieve high bit rates. Coding in this case is essential for achieving the highest possible rates in the presence of crosstalk and impulsive and other interference.

Proper coding design is extremely important for a digital communication link [1]. A designer should take several design factors into account. Those include required coding gain for intended link budget, channel characteristics, source coding requirements, modulation, etc. Coding in OFDM systems has an additional dimension. It can be implemented in time and frequency domain such that both dimensions are utilised to achieve better immunity against frequency and time selective fading. Obviously, interleaving plays an important role to achieve the above goal as shown in Figure 7.1. A combination of block coding and convolutional coding along with proper time/frequency interleaving constitutes a concatenated coding strategy. A new generation of parallel concatenated coding, namely turbo coding, also seems promising for some OFDM applications.

In this chapter, performance of block, convolutional, and trellis coding, and issues concerning concatenated and turbo coding are discussed. A comparison of their performance with theoretical bounds is also presented.

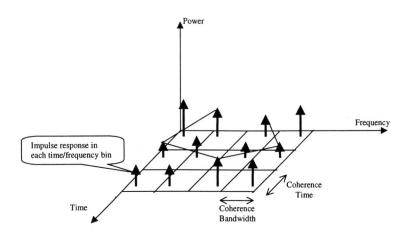

Figure 7.1. Two-dimensional coding for OFDM.

7.2   Block Coding in OFDM

In classical block coding, input data are blocked into groups of k bits, and each block is mapped into an output block of n bits, where $n > k$. In the canonical form, $n - k$ parity checks are computed among the input bits according to some algebraic procedure, and then appended to the original block. This is illustrated in Figure 7.2 This requires an increase in bandwidth by a factor of n/k. The reciprocal of this factor is the efficiency, or rate, of the code.

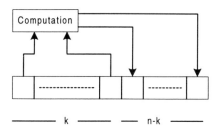

Figure 7.2. Construction of a canonical block code.

Only 2^k of the possible 2^n output blocks are legitimate code words. The code is chosen such that the minimum "Hamming distance", which is the number of bits in which code words differ, is maximized. The code is described by the set of numbers $[n, k, d]$, where d is the minimum Hamming distance.

At the receiver, the n–bit block is recovered, possibly with errors, by demodulation and framing. The decoder finds the permissible code word that is closest in Hamming distance to this received block. The $n - k$ check bits may then be deleted and the result output as a replica of the original input. If $d = 2t + 1$, then any set of t or fewer errors in the block can be corrected. In OFDM, if n agrees with the number of bits in an OFDM symbol, then each symbol is treated separately and no memory beyond a symbol is required for decoding.

The performance improvement achieved by coding is illustrated in Figure 7.3 for the Gaussian channel. Not only is the error probability substantially reduced, but the shape of the curve changed to one in which the error probability decreases very sharply with small increase in signal-to-noise ratio. For a relatively large block size, the sharp decrease occurs at a signal-to-noise ratio only a few dB higher than that which would yield an ideal capacity of the same bit rate. The sharpness of the curve implies that when such channel coding is used, little further benefit can be achieved by source coding that reduces the effects of errors.

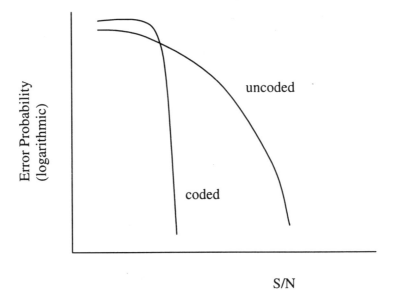

Figure 7.3. Performance with block coding over a Gaussian channel.

The decoding described above is "hard decision", that is the decoder operates on the binary output of the demodulator. The actual size of error in the analog domain in the demodulator is lost. Better performance could be achieved if a measure of that analog error were used in the decoding operation. This is referred to as "soft decoding". Its complexity is clearly much greater than hard decoding.

An intermediate approach between hard and soft decoding is the introduction of erasures. In this technique, the demodulator not only detects the coded bits, but also measures the reliability of those decisions. When the reliability is below some threshold, the demodulator outputs an erasure symbol instead of a bit decision. The decoder now operates on 3–level inputs. Any combination of t errors and e erasures can be corrected if $d = 2t + e + 1$. Approximately half of the performance difference between hard and soft decision decoding can be recovered over a Gaussian channel if the erasure threshold is optimized.

Block codes can operate over symbol alphabets higher than binary. A particularly powerful and widely used family of codes is the Reed-Solomon [2]. The alphabet size is any of the form $m = q^p$ where q is a prime number, almost always 2 in practice. An *[n, k, d]* code maps k m-ary symbols into n m-ary symbols. d is a generalized Hamming distance, the minimum number of symbols that differ between code words. The block size n must be less than or equal to m–*1*. A shortened block size leads to simpler implementation at the cost of performance.

As an example, consider a Reed-Solomon code that treats 8-bit sequences (bytes) as the underlying alphabet. Then $m = 2^8 = 256$, so that $n \le 255$. A {255, 235, 20] code maps 235-byte (1880 bits) blocks into 255-byte (2040 bits) blocks. It can correct up to 10 errors or 20 erasures, or an intermediate combination of both.

Block codes, in particular the Reed-Solomon class, are effective in combating burst errors. In single carrier systems, such bursts result from impulsive noise of duration greater than a symbol period, and from correlation due to fading when the time coherence of the channel is longer than a symbol period. The same considerations apply for a multi-carrier system, but in addition their frequency duals must be considered. An impulse will almost always have wide frequency content so as to affect several sub-carriers. A frequency selective fade whose coherence bandwidth is wider than the sub-carrier spacing will lead to correlated errors among the sub-carriers.

The performance of codes in the presence of bursts can be improved by the process of interleaving. Rather than operating on symbols that are adjacent in time and/or frequency, the code is made to operate on symbols with sufficient spacing so that the errors are more independent. A simple and common form of interleaving is periodic interleaving illustrated in Figure 7.4. At the receiver, the symbols are de-interleaved before decoding. The decoder therefore operates on symbols spaced m symbol periods apart as transmitted. The spacing can be in time, frequency, or both.

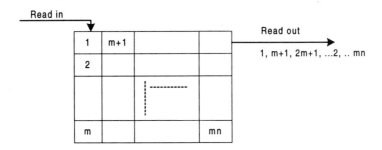

Figure 7.4. Implementation of periodic interleaving.

A simple form of block coding that is very widely used for error detection, not correction, is the cyclic redundancy check (CRC). Here a fixed number of check bits are appended to a block of arbitrary length. If the receiver detects any errors, a retransmission is requested. The encoding is based on an n-bit feedback shift register whose connection pattern is a primitive polynomial. The receiver employs a shift register with the same connection polynomial. This is shown below.

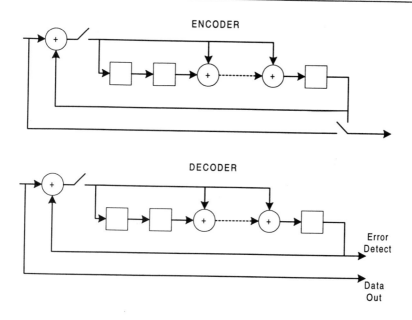

Figure 7.5. CRC implementation.

The operations shown are modulo-2 addition (exclusive-or). At the encoder, the shift register is initialized to a given pattern, such as all ones. The input data is fed both to the channel and to the feedback shift register. After the input block is completed, the contents of the shift register are shifted out to the channel. At the receiver, the shift register is also initialized. The received data is simultaneously output and fed to the register. After completion of the uncoded bits, the contents of the register are examined. An all-zero condition of these bits indicates that no errors have occurred. A CRC of length n can detect any error pattern of length n or less. Any other error pattern is detected with probability $1 - 2^{-n}$.

7.3   Convolutional Encoding

Another very important form of coding is convolutional. Instead of operating on symbols arranged as blocks, the coding operates continuously on streams of symbols. At the encoder, the input is fed continually through a shift register of length m. The memory of the code is the "constraint length" $(m+1)$, the number of output symbols affected by an input symbol. Each time a bit is read into the register, several modulo-2 sums of the present and past bits are formed. The choice of which bits are operated on is designated as a polynomial $P(z)$ with binary coefficients. n such modulo-2 sums are formed and multiplexed to form the output of the "mother code". Since n bits are generated for each input bit, the rate of the code is $1/n$. In an elaboration not treated here, more than one bit at a time is read in to a corresponding number of shift registers.

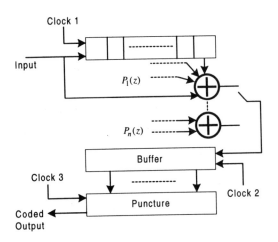

Figure 7.6. Generation of a convolutional code.

The rate of the code can be increased by the process of puncturing. This involves deleting some of the bits generated by the mother code. The process of generating a convolutional code is shown in Figure 7.6. In addition. interleaving may be applied to a convolutional code in a manner similar to a block code.

A convolutional code can be described by a state diagram. For a binary convolutional code, the number of states is $s = 2^m$, where m is the number of shift register stages in the transmitter. In the state diagram shown in Figure 7.7, the transitions from a current state to a next state are determined by the current input bit, and each transition produces an output of n bits.

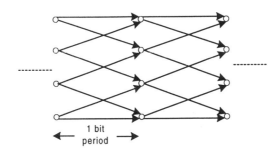

Figure 7.7. State diagram of a convolutional code.

A particular advantage of convolutional codes is the relative ease of decoding. Sequence estimation techniques are applicable, in particular the Viterbi algorithm. The Viterbi algorithm has the desirable properties of not only very efficient computation, but also a fixed amount of computation per symbol and an orderly flow of computation. The low complexity of decoding permits soft decision decoding, where the minimum Euclidean distance between the received sequence and all allowed sequences, rather than Hamming distance, is used to form decisions.

The Viterbi algorithm is a general dynamic programming approach for solving problems such as finding shortest paths. For sequence estimation as in convolutional decoding, the algorithm finds the transmitted sequence whose Euclidean distance, or equivalently whose accumulated square distance, is closest to the received sequence. The operation consists of finding the path of states, where a state denotes the past history of the sequence over the constraint length as described above.

Figure 7.8 illustrates an example of a Viterbi algorithm applied to a simple code with $m=2$, or $s=4$ states. Because of the structure of the code, only some of the transitions between adjacent states are possible. These transitions are determined by the input bit sequence. The allowed sequences are described by the sequence diagram shown below. Each transition corresponds to a received sequence of n bits, for which a tentative decision on one bit is performed.

The Viterbi algorithm at any instant keeps track of s survivor sequences up to that time, one terminating in each state, and an error metric associated with each such sequence. For a Gaussian channel, an optimum metric is the accumulated square of the distance between all of the bits in that sequence and the received sequence. One of those sequences is assumed to be correct, but which one is undecided.

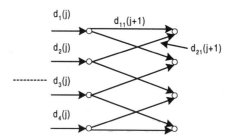

Figure 7.8. Viterbi algorithm applied to a simple convolutional code.

When a group of n bits corresponding to one information bit is received, each survivor sequence is extended according to the added metrics associated with the new input. Each possible sequence terminating in each new state is evaluated, and all sequences other than the one with minimum metric is discarded.

The figure shows how the survivor sequence for state 1 is chosen. The same process is performed for each state. For $r(k)$ the received signal over time k and $t(k)$ the transmitted symbols, then

$$d(j) = \sum_{i=0}^{j} |r(i) - t(i)|^2 \qquad\qquad 7.1$$

$$d_{pq}(j) = |r(j) - t_{pq}(j)|^2, \qquad\qquad 7.2$$

where $t_{pq}(j)$ is the symbol sequence associated with the transition from state p to state q. The values of r and t may be complex. The survivor sequence chosen for state 1 at time $j + 1$ is the one whose accumulated metric is

$$d_1(j+1) = \min\{d_p(j) + d_{pq}(j)\}. \qquad\qquad 7.3$$

Only allowed paths pq are examined. Then at time $j+1$ we again have one survivor sequence terminating in each state, each one being an extension of a previous survivor sequence. It is necessary to make a decision at each step. In almost all cases, if we project far enough into the past, all survivor sequences will have a common beginning. In practice going back $N = 6(m+1)$ is sufficient. So at any time the stored survivor sequences are N information symbols long, one past information symbol is read out, and the delay through the decoder is N.

The squared error is optimum for a Gaussian channel, but a sub-optimum metric for a fading channel. Nevertheless it is frequently used because of its simplicity. For optimum performance over a fading channel, the signal-to-noise ratio must somehow be measured for each received symbol. This measure is often referred to as the "channel state information". The squared error metric is then weighted for each symbol by this measure. Note that this is same form of weighting that is used in maximum ratio combining when diversity is employed. An intermediate approach is similar to the use of erasure symbols described previously. In this case the squared error is used whenever the signal-to-noise is above a properly chosen threshold. When the measured signal-to-noise is lower than the threshold, zero weighting is applied to the symbol, so that the accumulated metrics are not increased. Most of the loss due to the original sub-optimum metric is recovered.

For optimum operation of the Viterbi algorithm, the received inputs should be independent. This is approximated by sufficient interleaving, where the interleaving must effectively eliminate correlation in both time and frequency. For a sufficiently long constraint length and interleaving, a Rayleigh channel is converted to an approximately Gaussian one [3, 4]. The interleaving may be considered to be a form of diversity. Figure 7.9 illustrates the performance of an interleaved convolutional code over a Rayleigh channel. The interleaving in effect converts the channel to a Gaussian one, with a far lower error rate and steeper curve. The coding further improves and steepens the performance curve, in a manner previously sketched in Figure 7.3. This curve assumes ideal decoding using channel state information. For sub-optimum decoding, the curve will be less steep.

It is possible for the input and output of the convolutional encoder to be a continuous stream, with start-up only upon system initialization. However it is more common for the code to restart and terminate over some interval. That interval could in fact correspond to an OFDM symbol.

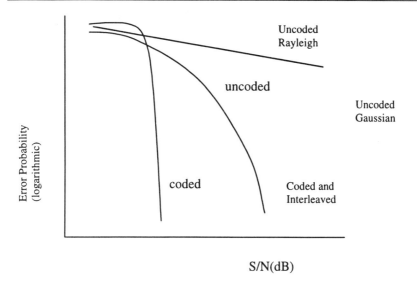

Figure 7.9. Performance of a convolutional code over a Rayleigh fading
channel.

In this case the encoder and decoder start in a given state and are forced
to terminate in a given state. This allows the Viterbi decoder to eliminate all
survivor states but one at the end and to read out the remainder of the stored
detected sequence.

7.4 Concatenated Coding

Combining convolutional and block codes in a concatenated code is a
particularly powerful technique. The block code is the outer code, that is it is
applied first at the transmitter and last at the receiver. The inner
convolutional code is very effective at reducing the error probability,

particularly when soft decision decoding is employed. This operation is shown in Figure 7.10. However when a convolutional code does make an error, it appears as a large burst. This occurs when the Viterbi algorithm chooses a wrong sequence. The outer block code, especially an interleaved Reed-Solomon code, is then very effective in correcting that burst error. For maximum effectiveness the two codes should be interleaved, with different interleaving patterns. Both interleaving patterns should account for both time and frequency correlation. Reference [3] evaluates the performance of several combinations of convolutional and Reed-Solomon codes for an OFDM system operating over a fading channel.

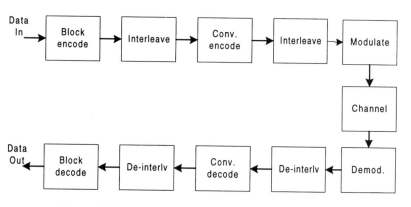

Figure 7.10. Concatenated coding with interleaving.

7.5   Trellis Coding in OFDM

Trellis coding is similar in spirit to convolutional coding, but differs in the very important aspect that the coding is imbedded in the modulation rather than existing as a separable process [5, 6]. Soft decision decoding is based on minimum Euclidean distance of sequences, and is part of the demodulation procedure.

This is in keeping with the basic philosophy of information theory. In Shannon's original paper, capacity is achieved by assigning a different waveform for each complete message, and the length of the message is allowed to approach infinity. Nowhere is the message described as a sequence of bits or symbols. The receiver chooses the message as the one closest to the received waveform. Trellis coding is a practical approximation to this theoretical scheme.

Rather than adding redundant bits or symbols prior to modulation, redundancy is introduced by using constellations with more points than would be required without coding. Usually the number of points is doubled. The number of symbols per second is unchanged, therefore the bandwidth required is also unchanged. Since there are more possible points per symbol, it may appear that the error probability for a given signal-to-noise ratio would increase. However, as in convolutional coding, dependencies are introduced among the different symbols in that only certain sequences of constellation points are allowed. By properly making use of these constraints in the receiver, as in convolutional decoding, the error probability actually decreases. A measure of performance improvement is the "coding gain", which is the difference in S/N between a coded and an uncoded system of the same information rate that produces a given error probability.

The first step in designing a trellis code is to form an expanded constellation and to partition it into subsets. The points within each subset are made far apart in Euclidean distance, and will correspond to uncoded bits. The remaining, or coded bits, determine the choice of subset. Only certain sequences of subsets are permitted, those sequences being determined by a simple convolutional code. As in convolutional coding, the sequence is best described by a state transition diagram or " trellis". In order to keep allowed sequences far apart, the constellation subsets are chosen so that those corresponding to branching in and out of each state have maximum distance separation.

We will illustrate the procedure with a simple trellis code that uses a 16-point QAM constellation to carry 3 bits per symbol. The technique is also applicable to phase modulation, where the number of phases is increased. Shown in Figure 7.11 is a partition of the constellation into 4 subsets of 4 points each, where the labelling of points denotes subset membership.

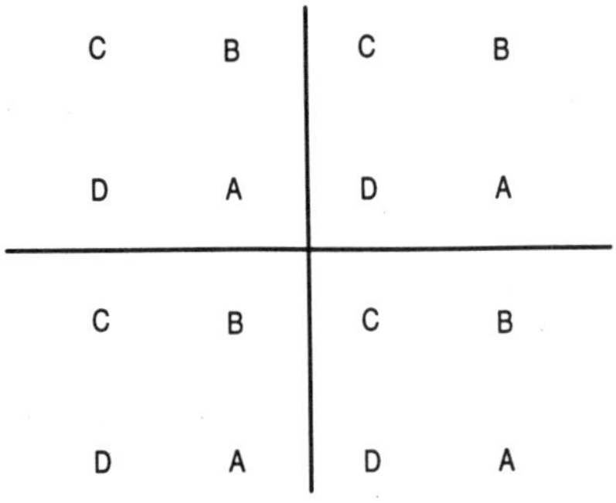

C	B	C	B
D	A	D	A
C	B	C	B
D	A	D	A

Figure 7.11. An expanded constellation partitioned for Trellis coding.

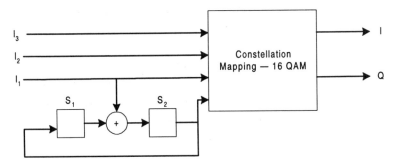

Figure 7.12. Implementation of a Trellis code.

Figure 7.12 shows the mapping of 3-bit input symbols into constellation points. Two of the input bits, I_2 and I_3, are uncoded and determine which member of a subset is used. The third bit I_1 is fed to a rate-1/2 convolutional encoder whose 2-bit output determines which subset is used.

Figure 7.13 is the state diagram of the code. The states are labelled according to the contents of the shift register in the convolutional encoder. The labels of the transitions indicate which subset is used when that transition occurs.

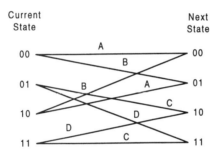

Figure 7.13. State diagram of the Trellis code.

At the receiver, a Viterbi algorithm is used for the combined demodulation and decoding. For each received symbol the distance to the nearest member of each subset is measured, as shown in Figure 7.14. The square of this value serves as the metric in extending the survivor states of the Viterbi algorithm. For each survivor state, not only must the coded bits and accumulated squared distance be stored, but also the uncoded bits corresponding to which member of the subset was the nearest point for each symbol.

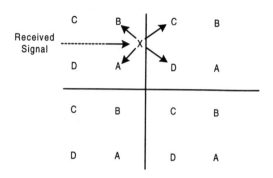

Figure 7.14. Metrics used for Viterbi decoding.

The code described above operates on 2-dimensional symbols. Doubling the number of points reduces the distance between points by a factor of $\frac{1}{\sqrt{2}}$, so an initial loss of 3 dB (for a large constellation) must be more than compensated for. Performance can be improved by operating on symbols of more than 2-dimensions. For example, pairs of QAM symbols may be treated as 4-dimensional symbols. By doubling the number of points in 4-dimensions, spacing between points now reduced by $\frac{1}{\sqrt[4]{2}}$, for an initial loss of 1.5 dB.

The error probability curve over a Gaussian channel is approximately parallel to the uncoded curve, shifted by the coding gain. Coding gains of up to 6 dB are theoretically possible, with up to 5 dB achieved by practical trellis codes. The very simple 4-state, 2-dimensional code described above provides approximately 3 dB coding gain. If for the same constellation we had used an 8-state code with 1 uncoded bit and 2 coded bits, then the coding gain would be increased to approximately 4 dB.

In OFDM it is the usual practice to perform trellis coding over the sub-carriers of a single OFDM symbol, rather than extending memory over a greater interval of time. At the beginning of each OFDM symbol, the code is started in a known state. The code is forced to a known state at the end of an OFDM symbol. Any null sub-carriers are skipped. Variable size constellations are readily handled by using a fixed number of coded bits on each sub-carrier and a variable number of uncoded bits.

7.6   Turbo Coding in OFDM

Turbo coding has recently been used in many communication systems successfully. Random-like structure of turbo codes has resulted in outstanding performance by providing small error rates at information rates of close to theoretical channel capacity. In this section, we briefly review some fundamental features of turbo coding and explain why it is a good potential candidate for OFDM systems.

In general a turbo code comprises two or more concatenated or parallel codes. A typical turbo code includes parallel concatenated convolutional codes where the information bits are coded by two or more recursive systematic convolutional codes each applied to permutations of the information sequence as in Figure 7.15. However, the same iterative turbo decoding principle can be applied to serially and hybrid concatenated codes [7]. The high error correction power of turbo coders originates from random like coding achieved by random interleaving in conjunction with concatenated coding and iterative decoding using (almost) uncorrelated extrinsic information.

The structure and complexity of turbo encoder design is restricted by other system parameters. Here, we briefly review some of the critical issues which directly impact the code structure. *Decoding Delay* is critical for receiver design. Since the decoding structure of turbo codes is iterative and includes an interleaving /de-interleaving block for each iteration, the excess delay may interfere with overall system performance. For example, speech frames should be processed within a tight time frame in most applications. Another important system design issue is required *coding gain*. While the BER improvement of turbo coders is outstanding, a system designer should be aware of the consequences of low SNR system functionality. It is tempting to design the system for a low SNR condition assuming that a powerful turbo code can meet the target BER performance. However, many other receiver functions such as synchronization and adaptive algorithms require a minimum SNR which should be taken into account carefully.

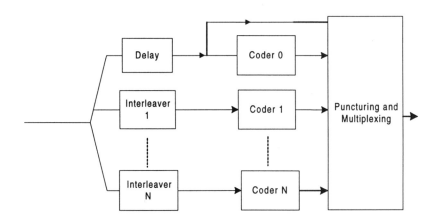

Figure 7.15. A typical turbo encoder.

The decoding process is iterative and in this overview we briefly explain two common iterative techniques: Maximum *a posteriori* Probability (MAP) technique and Soft input Soft Output Viterbi Algorithm (SOVA). Three different types of soft inputs are available for each decoder: the uncoded information symbols, the redundant information resulting from first Recursive Symmetric Code (RSC) and *a priori* (extrinsic) information. The output is a weighted version of uncoded information, a priori information and new extrinsic information.

In general, a symbol-by-symbol MAP algorithm is optimal for state estimation of a Markov process. MAP algorithms for turbo decoding calculate the logarithm of the ratio of *a posteriori* probability (APP) of each information bit being one to the *a posteriori* probability of the bit being zero. The MAP technique is complicated and requires non-linear operations which makes it less attractive for practical purposes. In comparison, a simplification of MAP algorithm, namely, Soft Output Viterbi Algorithm (SOVA) leads to a practical sub-optimum technique. While performance is slightly inferior to an optimal MAP algorithm, the complexity is significantly less. MAP takes into account all paths by splitting them into two sets, namely, the path that has an information bit one at a particular step and paths which have bit zero at that step and returns the log likelihood ratio of the two. SOVA considers only the survivor path of the Viterbi algorithm. Therefore, only the survived competing path which joins the path chosen in the Viterbi algorithm is taken into account for reliability estimation.

While the encoders have a parallel structure, the decoders operate in serial which results in an asymmetric overall structure. For example, in the first iteration the first decoder has no *a priori* information [9].

As discussed before, coding schemes for OFDM systems take advantages of coding in both time and frequency in conjunction with proper interleaving. This provides extra protection against time and frequency selective fading. Use of turbo coding for OFDM systems is seriously being considered for several wireless standards. The depth of interleaving can be restricted to one OFDM block which provides correlation between symbols modulated by different sub-carriers and subjected to independent selective fading.

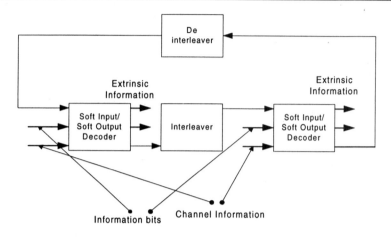

Figure 7.16. A turbo decoder structure.

Another coding scheme supports inter-block interleaving which can be intelligently designed to provide time and frequency correlation. While the former technique does not impose extra delay at the receiver, the latter technique gives another degree of freedom to the system designer. The choice of proper coding scheme depends on system requirements and applications.

References

1. Costello, D.J., Costello Lin. *Error Control Coding.* Englewood Cliffs, NJ: Prentice-Hall, 1983.

2. Wicker, B., Bhargava V.K. *Reed-Solomon Codes and Their Applications.* New York: IEEE Press, 1994.

3. Alard M., Lassalle R. "Principles of Modulation and Channel Coding for Digital Broadcasting for Mobile Receivers." *EBU Review*; Aug 1987; 224: 47-69.

4. Pommier D., Wu Y. "Interleaving or Spectrum Spreading in Digital Radio Intended for Vehicles." *EBU Review*; Jun 1986; 217: 31-44.

5. Ungerboeck, G. "Trellis Coded Modulation with Redundant Signal Sets." *IEEE Commun. Mag.*; Feb 1987; 25; 2: 5-21.

6. Biglieri, D., Divsalar D., McLane P.J., Simon M.K. *Introduction to Trellis-Coded Modulation with Applications.* New York: Macmillan, 1991.

7. Hagenauer, J., Offer E., Papke L. "Iterative Decoding of Binary Block and Convolutional Codes." *IEEE Trans. Information Theory* ;Feb 1987; 25; 2: 5-21.

8. Jung, P. "Comparison of Turbo-Code Decoders Applied to Short Frame Transmission Systems." *IEEE Journal on Selected Areas in Communications;* Apr 1996; 14; 3.

9. Hoeher, P. "New Iterative Turbo Decoding Algorithms" *Proceeding of The International Symposium on Turbo Codes*; Brest, France; 1997.

Chapter 8 *ADSL*

8.1 Wired Access to High Rate Digital Services

A ubiquitous communication channel is the subscriber line, or loop, consisting of an unshielded twisted pair of wires, connecting any home or office to a telephone company's central office. The overwhelming majority of these channels are used to carry analog voice conversations, which require a bandwidth of less than 4 kHz. It has long been recognized that most subscriber lines can support a much wider bandwidth, in particular, to carry high rate digital signals. The first such widespread use is for access to basic rate ISDN, in which the subscriber line carries 160 kb/s simultaneously in both directions over a single pair.

More recently, higher rates have been introduced into numerous systems. Of particular interest here is ADSL which is primarily intended to provide access for residential applications [1, 2]. Most of such applications require a high data rate in the downstream direction (to the customer) and a much lower rate upstream (from the customer). The primary applications of ADSL are the delivery of digitally encoded video, and access to digital

159

services, particularly the Internet. ADSL meets these needs by providing a high rate digital downstream signal of over 1 Mb/s, a moderate rate upstream signal, and a normal analog voice channel, all over a single wire-pair. Because virtually all customers have a wire-pair channel providing voice service, no additional channel need be installed to provide this new service. It only requires the installation of terminating equipment at the customers' premises and at the central offices. OFDM, typically referred to as DMT (Discrete Multi-Tone) in this application, has been adopted as the standard for transmission of the digital information. Several varieties of ADSL exist, differing primarily in the bit rates carried and the set of subscriber lines over which satisfactory performance is achieved.

8.2 Properties of the Wire-Pair Channel

Before discussing the transmission techniques, it is essential to first describe the channel — both its transmission properties and the types of interference to which it is subject. As with any communication channel, a combination of both factors must be considered in determining the performance that can be achieved.

The connection between the subscriber and the central office consists of several sections, each consisting of a wire-pair bundled in a cable with many other wire-pairs [3, 4, 5]. Insulation on modern cables is polyethylene, although older paper pulp cable is still present in much of the plant. The sections nearest the central office are referred to as feeder, where the cable size is large but the wire gauge may be fine. Closer to the customer, the distribution plant consists of smaller cables, but typically heavier gauge wire. Finally, the individual drop wire provides the final connection to the subscriber. Wiring on premises may also be significant. Typical wire gauge size throughout is 24-gauge (0.5 mm) or 26-gauge (0.4 mm). The overall length of the circuit can be up to 5.5 km. If it is greater than this distance, loading coils are usually added which improve transmission in the voiceband but greatly restrict the bandwidth above 4 kHz. Loaded circuits therefore cannot be used to provide any high rate digital service. Fortunately such circuits are rare and exist only in rural areas.

A trend is now taking place to introduce carrier systems and remote switches into subscriber plants. Such electronic equipment is connected to the central office digitally, often using optical fiber. ADSL in this case need only provide access from the subscriber to the remote electronic device rather than all the way to the central office. A plan, referred to as the Carrier Serving Area (CSA), envisions almost all future subscribers within 3.7 km of either a central office or of remote equipment if the cable used for the whole path is 0.5 mm or larger, or 2.7 km if 0.4 mm wire is used. An additional benefit of this plan is the elimination of loaded circuits.

The transmission properties of a wire-pair are determined by its primary constants, the resistance R, inductance L, capacitance C, and conductance G per unit length. At low frequencies, R is constant, but increases proportional to $\sqrt{f}$ at high frequencies (above about 200 kHz), due to skin effect. C is very constant with frequency, and L declines slightly with frequency. The conductance increases with frequency, but is small enough for modern cables that it may be neglected over the entire frequency range. Ignoring mismatched impedance effects, the transfer function of a line of length d is

$$H(f) = e^{-\gamma(f)d},$$

8.1

where the complex propagation constant γ is given by

$$\gamma = \alpha + j\beta = \sqrt{(R + j\omega L)(G + j\omega C)}.$$

8.2

α is the loss in nepers per unit length, and β the phase shift in radians per unit length. The phase distortion can, in general, be equalized, at least in principle. So, the loss has the primary effect on performance. However, it should be noted that the phase distortion, that is the deviation from linear phase vs. frequency (flat delay), is very severe at low frequencies. At low frequencies

$$\alpha \approx \sqrt{\frac{\omega RC}{2}}$$

8.3

and is therefore proportional to $\sqrt{f}$.

At high frequencies

$$\alpha \approx \frac{R}{2}\sqrt{\frac{C}{L}}$$

8.4

so the loss is also proportional to $\sqrt{f}$ because of skin effect on R and the relative constancy of L.

In between, roughly in the range of 20 – 200 kHz, the loss increases more slowly with frequency. Figure 8.1 is a plot of loss vs. frequency for a typical cable pair. Note the slope of ½ at the upper and lower ends, and the lower slope in the middle range.

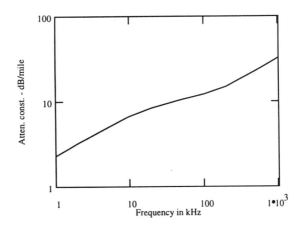

Figure 8.1. Attenuation constant of 24-gauge (0.5 mm) wire-pair.

A further consideration affecting the transfer function of a wire-pair is mismatched termination which results in reflection. Impedance matching over all frequencies is difficult, particularly at low frequencies where the characteristic impedance Z_0 of the line is highly variable, becoming infinite as $f \to 0$.

Another effect to be considered is the presence of bridged taps, which exist in many parts of the world, including the USA, but are not present in some other countries. Bridged taps are unused sections of line attached to the line of interest, resulting from the telephone company's loop installation and removal policies. A bridged tap is open circuit at its end, and therefore has no effect at low frequency such as the voice band. However at high frequency, in particular where the length of the tap is near ¼ wavelength, the tap presents a low impedance shunt and reduces the transfer function. In addition, reflection from the end of the tap is a source of echo in both directions.

Figure 8.2 illustrates a bridged tap and its effect:

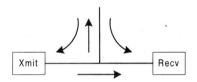

Figure 8.2. A bridged tap and its echoes.

Overall, unlike a radio or voiceband channel, the wire-pair channel is not sharply bandlimited but has a loss that increases gradually with frequency. This brings up the non-trivial question of the effective bandwidth of such a channel. It would be a mistake to consider only the loss in answering this question. The usable bandwidth depends equally on the noise present, because if the noise is low the usable bandwidth can be quite high, even though the loss at high frequency is high.

The primary source of noise is crosstalk from signals on other pairs in the cable [6, 7]. As illustrated in Figure 8.3, near-end crosstalk (NEXT) results from transmit sources co-located or close to the receiver in question, while far-end crosstalk (FEXT) is caused by sources at the other end of the channel. When both effects are present, NEXT is more severe because of the relative signal levels. NEXT may expected to be a more serious problem at the central office rather than at the customer end because many co-located signals are more likely to occur.

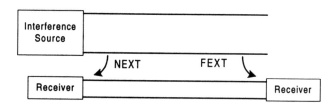

Figure 8.3. Crosstalk mechanism.

Crosstalk results both from capacitive and inductive coupling between wire-pairs in the same cable. Figure 8.4 illustrates the capacitive coupling mechanism.

Shown is a differential length of cable in which a signal on the upper pair crosstalks into the lower pair. The four capacitances between wires of one pair and wires of the other pair are shown. As shown, there are two current paths of opposite sign.

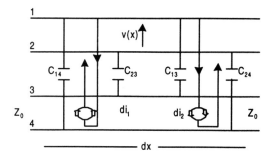

Figure 8.4. Capacitive crosstalk mechanism.

The net differential current is

$$di_c = di_1 - di_2 = \cfrac{V(x)}{\cfrac{1}{j\omega C_{13}dx} + \cfrac{1}{j\omega C_{24}dx}} - \cfrac{V(x)}{\cfrac{1}{j\omega C_{14}dx} + \cfrac{1}{j\omega C_{23}dx}}$$

$$= j\omega V(x)\left[\frac{C_{13}C_{24}}{C_{13} + C_{24}} - \frac{C_{14}C_{23}}{C_{14} + C_{23}}\right]dx.$$

8.5

If the differences between capacitances are much less than the capacitances themselves, as is almost always true, then the above quantity may be approximated by

$$di_c = \frac{j\omega C_u}{4}V(x)dx,$$

8.6

where C_u is an unbalance capacitance given by

$$C_u = C_{13} - C_{14} + C_{24} - C_{23}.$$

8.7

This quantity can be either positive or negative, and varies along the line, particularly as the relative twist between the two pairs changes. Over the length of the line, d, it will be zero mean. The induced incremental voltage is

$$dv_c = \frac{Z_0}{2} di_c = \frac{j\omega V(x)C_u Z_0}{8} dx, \qquad 8.8$$

where it is assumed that the line is terminated in its characteristic impedance in both directions.

The inductive coupling may be treated similarly, as shown in Figure 8.5.

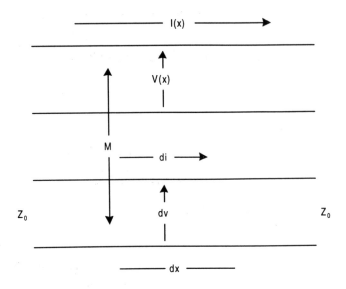

Figure 8.5. Inductive crosstalk mechanism.

Over the differential length, the mutual inductance is $M\,dx$. Then the differential induced voltage is

$$dv_m = j\omega I(x)Mdx, \text{ and } di_m = \frac{j\omega I(x)Mdx}{2Z_0}. \qquad 8.9$$

The near-end and far-end currents are

$$di_n = di_c + di_m, \text{ and } di_f = di_c - di_m . \qquad 8.10$$

These can be expressed as

$$di_n = j\omega Y_n dx, \text{ and } di_f = j\omega Y_f dx, \qquad 8.11$$

where Y_n and Y_f are unbalanced functions given by

$$\frac{C_u Z_0}{8} \pm \frac{M}{2Z_0}. \qquad 8.12$$

They are both zero mean with the same variance, and non-correlated over distances comparable to a twist length. Except at low frequency where Z_0 is highly variable, they are approximately constant with frequency.

We are now ready to evaluate overall crosstalk. Figure 8.6 illustrates the case of NEXT.

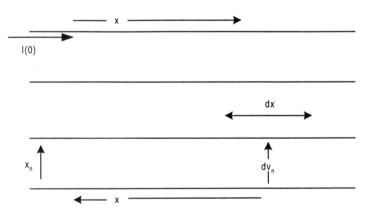

Figure 8.6. Near-end crosstalk generation.

If the interfering source current is denoted by *I(0)*, then at a distance of x it has been attenuated to $I(0)e^{-\gamma x}$. The incremental induced crosstalk current is then

$$di_n(x) = j\omega Y_n(x) I(0)\ e^{-\gamma x}dx,$$

8.13

and similarly, the incremental crosstalk voltage is

$$dv_n(x) = j\omega Y_n(x) V(0)\ e^{-\gamma x}dx.$$

8.14

When propagated back to the input, the contribution of this crosstalk component is

$$dx_n = dv_n(x)e^{-\gamma x}dx = j\omega Y_n(x)V(0)e^{-2\gamma x}dx$$

8.15

and the total NEXT voltage is

$$x_n = j\omega V(0)\int_0^d Y_n(x)\ e^{-2\gamma x}dx.$$

8.16

This quantity is zero mean, with variance

$$\overline{x_n^2} = \omega^2 V^2(0)\int_0^d\int_0^d Y_n(x)Y_n(y)\ e^{-2\gamma x}e^{-2\gamma^* x}dxdy.$$

8.17

Because Y_n is non-correlated over short distances, we can approximate

$$< Y_n(x)Y_n(y) > = k_n\delta(x-y)$$

8.18

so that

$$\overline{x_n^2} = \omega^2 V^2(0)k_n\int_0^d e^{-4\alpha x}dx = \frac{\omega^2 V^2(0)k_n}{4\alpha}(1-e^{-4\alpha d}).$$

8.19

For any but the shortest lines, $e^{-4\alpha d} \ll 1$. We will further approximate $\alpha \approx k\sqrt{f}$. Then, lumping together the various constants,

$$\overline{x_n^2} \approx V^2(0)\lambda_n f^{3/2}.$$

8.20

We now have the important result that the NEXT power transfer function,

$$T_n(f) = \frac{\overline{x_n^2}}{V^2(0)} \qquad\qquad 8.21$$

is proportional to $f^{3/2}$, and is independent of the line length except for very short lines. The level of the transfer function is determined by the constant λ_n. It can also be specified by giving its value at some frequency f_0. Then at any other frequency,

$$T_n(f) = T_n(f_0) + 15\log_{10}\frac{f}{f_0} \quad (\text{ in dB}). \qquad\qquad 8.22$$

We can analyze the far-end crosstalk in a similar manner to obtain

$$T_f(f) = \frac{\overline{x_f^2}}{V^2(0)} = \lambda_f f^2 d\, e^{-2\alpha(f)d} \qquad\qquad 8.23$$

The exponential factor arises from the fact that all crosstalk components travel a total distance of d, partly on the interfering pair, and the remainder on the interfered pair. When both NEXT and FEXT occur, the former dominates at all frequencies of interest.

It is usual to model various system imperfections, such as finite precision digital implementations, as an added white Gaussian noise, with spectral density N_0. This is usually a reasonably accurate model. In addition, it eliminates singularities in signal-to-noise ratios in the presence of only crosstalk. It should be noted that at the relatively high receiver power levels in wire-pair systems, actual thermal noise is so far below the signal that it becomes completely negligible.

We are now ready to express the signal-to-noise ratio, which is the principal determinant of performance. Let the transmitted signal's power spectral density be given as $G(f)$, and each of k interferers as $G_k(f)$. The desired channel power transfer function is $|H(f)|^2$, and that of the crosstalk paths is $T_k(f)$. Then

$$SNR(f) = \frac{G(f)\,|H(f)|^2}{\sum_k G_k(f)T_k(f) + N_0}.$$ 8.24

In an OFDM system the above quantity is an excellent approximation for the SNR of each sub-channel, where f is center frequency of the sub-channel.

The ideal capacity of the channel is

$$C = \int_0^{\infty} \log_2[1 + SNR(f)]df.$$ 8.25

A useful definition of the usable bandwidth of a wire-pair channel is the frequency range over which the SNR is greater than some value. For coded single carrier systems, a value of 0 dB is a reasonable one. This is also the range that typically includes over 95% of the ideal capacity. For an OFDM system, for each sub-carrier to carry at least a four-point constellation, the SNR should be at least approximately 10 dB.

8.3   ADSL Systems

Two classes of ADSL have been standardized recently, with many options in each. Full rate ADSL can carry up to approximately 8 Mb/s downstream and 800 kb/s upstream. A simpler class, commonly called "ADSL Lite," carries up to approximately 1.5 Mb/s downstream and 500 kb/s upstream. In both cases, data rates can be adjusted to any value in steps of 32 kb/s. An analog voice channel is provided on the same pair. The target error probability is 10^{-7} per bit, with some required margin [8, 9].

The two classes are somewhat compatible with each other. In both cases, sub-carriers are spaced 4312.5 Hz apart in both directions. After every 68 frames of data, a synchronization frame is inserted. Because of this and the use of cyclic prefixes, the net useful number of data frames is 4000 per second in all cases. One of the sub-carriers of the frame is devoted to synchronization. Adaptive bit allocation over the sub-carriers is performed in all cases. This process is critical to ensure system performance.

In the full rate downstream direction, a block of 255 complex data symbols, including several of value zero, are assembled. These will correspond to sub-channels 1 to 255. The lower ones cannot be used because of the analog voice channels, nor can the 255th. Therefore, the highest frequency allowed sub-carrier is centered at 1.095 MHz. Sub-carriers which cannot support at least a 4-point constellation at the desired error probability will also be unused. Conjugate appending is performed on the block, followed by a 512-point DFT. This results in frame of 512 real values. A cyclic prefix of 32 samples is added, and the resultant 2.208 M samples per second transmitted over the line.

Upstream, 31 sub-channels are processed, although (again) the lower few and the 31st cannot be used. The same processing is performed with a cyclic prefix of 4 samples. The upstream and downstream sub-channels may overlap. This provides a larger data rate, but requires the use of echo cancellation.

The bit streams may be treated as several multiplexed data channels. Each such channel may be optionally Reed-Solomon coded, with a choice of code and interleaving depth. Other optional codes include a CRC error check, and a 16-state 4-dimensional trellis code. The trellis code, when present, operates over the non-zero sub-carriers of a block, and is forced to terminate at the end of each block.

"ADSL Lite" is intended as a simpler lower cost system, with greater range of coverage because of the lower rate. One important difference is the elimination of filters at the customer's premises to separate the voice and data channels. The upstream channel is created identically to that of the full rate system, except that the first 6 sub-carriers must be zero.

The downstream transmitted sampled rate is reduced by a factor of two, to 1.104 M samples per second. The IDFT is performed over an initial block of 127 complex numbers, of which the first 32 must be zero. The highest sub-carrier is now at 543 kHz. In this case, the upstream and downstream sub-carriers do not overlap. The signal is treated as a single bit stream. Reed-Solomon and CRC coding are again optional, but there is no trellis coding.

As of the time of publication of this text, many trials of various ADSL systems are in progress around the world. The ultimate market for high rate digital residential services is not clear, nor are the relative advantages and disadvantages of providing these services over competitive cable television facilities.

References

1.   Starr, T., Cioffi J.M., Silverman P. *Understanding Digital Subscriber Line Technology,* Englewood Cliffs, NJ: Prentice Hall, 1999.

2.   Kyees, P.J., McConnell R.C., Sistanizadeh K. "ADSL: A New Twisted Pair Access to the Information Highway." *IEEE Commun. Mag.*; Apr 1995; 33: 52-59.

3.   Ahamed, S.V., Bohn P.P., Gottfried N.L. "A Tutorial on Two Wire Digital Transmission in the Loop Plant." *IEEE Trans. Commun.*; Nov 1981; 29: 1554-1564.

4.   Adams, P.F., Cook J.W. "A Review of Copper Pair Local Loop Transmission Systems." *Brit. Telecom. Tech. J.*; 1989; 7: 17-29.

5.   Werner, J.J. "The HDSL Environment." *IEEE J. on Sel. Areas in Commun.*; Aug 1991; SAC-9: 785-800.

6. Gibbs, J., Addie R. "The Covariance of Near End Crosstalk and its Application to PCM System Engineering in Multipair Cable." *IEEE Trans. Commun.*; Feb 1979; 27: 469-477.

7. Lin, S.H. "Statistical Behavior of Multipair Crosstalk." *Bell Sys. Tech. J.*; Jul-Aug 1980; 59: 955-974.

8. Chow, P.S., Tu J.C., Cioffi J.M. "A Discrete Multitone Transceiver for HDSL Applications." *IEEE J. on Sel. Areas in Commun.*; Aug 1991; SAC-9: 895-908.

9. Barton, M., Chang L., Hsing T.R. "Performance Study of High Speed Asymmetric Digital Subscriber Line Technology." *IEEE Trans. Commun.*; Feb 1996; 44: 156-157.

Chapter 9 *Wireless LAN*

9.1 Introduction

Growing interest in high speed wireless packet switched services such as wireless Internet access and wireless multimedia on one hand, and advances in integrated personal computers and communication devices on the other hand has resurrected high interest in wireless LANs technology. Wireless LAN are meant for low mobility, low delay spread and high data rate, such as indoor, environment.

Layering structure of Wireless LAN corresponds to lowest layers of ISO reference model of Open System Inter-Connection and is shown in Figure 9.1.

175

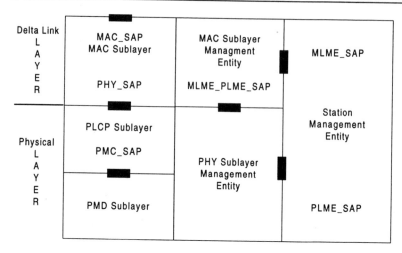

Figure 9.1. WLAN layers and corresponding ISO layers [1]

The MAC layer of wireless LAN networks should be designed to efficiently operate in an interference limited mobile environment. We review some of the fundamental differences between wired LANs and wireless networks. Hidden terminal effect where a terminal does not sense another terminal's transmission and consequently attempts to transmit is one source of packet collision. Unlike wired environment not all collisions result in loss of collided packets. Due to different transmit power of terminals and channel characteristics, a receiver may be able to correctly detect one of the collided packets. This phenomena is usually referred to as capture effect. Imperfect sensing due to multi-path fading experienced by received packets is another source of collision. Mobility of terminals require a handoff scheme and proper routing strategy.

Two different modes of operation for MAC layer is provisioned in the standard. The first mode is based on distributed coordination function (DCF) which is similar to best effort packet delivery techniques. The second mode of operation, point coordination function (PCF), is based on centrally controlled polling administered by access point (AP) and is used for delay sensitive data services.

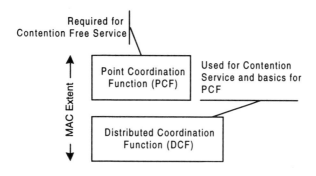

Figure 9.2. MAC architecture [1].

Access points are connected to distribution systems which could be Ethernet LAN, FDDI or other backbone infrastructure.

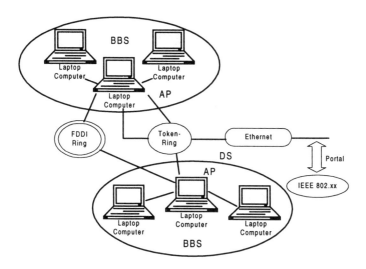

Figure 9.3. WLAN network architecture.

Support of DCF is mandatory for wireless LAN systems. An architecture which supports only DCF mode is called an ad-hoc network. In this mode all users have equal chance of using network resources.

The MAC sub-layer proposed for IEEE 802.11-compliant wireless LAN supports carrier sense multiple access with collision avoidance (CSMA/CA) algorithm similar to Ethernet's MAC layer. A wireless terminal can not listen to its own transmission because of hardware limitation, so CSMA/CD is not feasible. In CSMA/CA, a terminal senses the channel before packet transmission. If the channel is free for a period longer than distributed inter-frame spacing (DIFS) transmission occurs. Otherwise, it continues sensing until the channel becomes available. Collision avoidance feature requires the terminal to randomly delay the transmission upon channel availability for a DIFS period. The exponential random back-off is an integer number in time slot units. It should be noticed that unlike slotted Aloha, the slot time in this standard is much smaller than a packet length and is used for some timing coordination. In case of collision, a retransmission attempt requires a larger back-off time. Obviously there is no guaranteed minimum delay in DCF scheme.

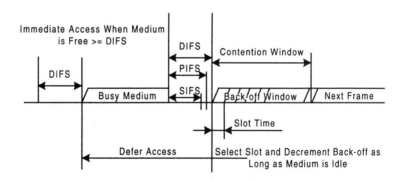

Figure 9.4. Distributed coordination function in WLAN [1].

For delay sensitive applications, such as voice over IP, the optional PCF provides a contention free (CF) window by a polling coordinator (PC) which polls stations and coordinates the transmission without contention. The basic operation of DCF and PCF is shown in Figure 9.4 and Figure 9.5.

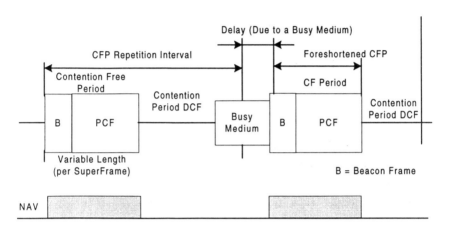

Figure 9.5. Point coordination function [1].

An important aspect of IEEE 802.11 is that a single common MAC layer is used for all physical layers. Therefore, interoperability among current and future physical layer techniques is feasible.

The physical layer consists of two sub-layers. Physical Layer Convergence Procedure (PLCP) which maps MAC sub-layer Protocol Data Units (MPDU) into a framing format appropriate for sending and receiving data between stations. Physical Medium Dependent (PMD) sub-layer defines the physical layer requirements for sending and receiving data through a wireless medium between stations. In this chapter we focus on PMD sub-layer functionality. For other sub-layers refer to [1]. Each frame comprises of headers for MAC layer and physical layer information as shown in Figure 9.6.

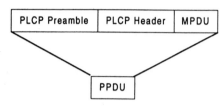

MPDU: MAC Protocol Data Unit
PPDU: PLCP Protocol Data Unit

Figure 9.6. Frame format for wireless LAN.

The PLCP preamble usually includes synchronization and frame delimiter fields. SYNC and Frame delimiter fields are used for adjusting receiver gain, detecting the packet, synchronization of bit clock and frame start, frequency offset estimation, timing recovery, and channel estimation to help PHY circuitry reach steady state demodulation stage.

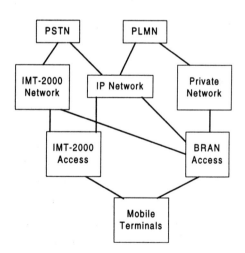

Figure 9.7. IMT2000 and WLAN convergence.

PLCP header contains information about physical link specifications, packet length and coding bits. A parallel converging line of activities in wireless LAN is conducted by ETSI under BRAN project.. Broadband wireless network architecture of BRAN includes HIPERLAN for data rates of up to 25 Mb/s and indoor applications, HIPERACCESS for wider area applications, and finally HIPERLINK for data rates of up to 155 Mb/s. Integration of BRAN radio access with IMT-2000 Network and IP networks is a noble sought after ongoing effort. A typical scenario is shown in Figure 9.7.

9.2 Physical Layer Techniques for Wireless LAN

Wireless LAN 802.11 supports a variety of physical media and communication techniques ranging from Frequency Hopping Spread Spectrum (FHSS), Direct Sequence Spread Spectrum (DSSS), OFDM to infrared light systems. Infrared is a low cost solution for low-range applications while radio is a better solution for larger areas with little line of sight access, it could be a proper complement to the radio part. Allocated spectra for wireless LAN applications are typically in 2.4 GHz and 5 GHz range where the bandwidth is scarce and much in demand. WLAN system at 2.4 GHz must meet ISM band requirements. The IEEE 802.11 standard supports both FHSS and DSSS techniques for this band. Spread spectrum techniques are robust against interference and do not require adaptive equalization. However, for higher data rates the synchronization requirements of spread spectrum techniques are more restrictive and add to system complexity. For data rates of above 10 Mb/s OFDM system shows better performance.

9.3   OFDM for Wireless LAN

Multi-carrier modulation is a strong candidate for packet switched wireless applications and offers several advantages over single carrier approaches. For higher data rate applications ranging from 10 Mb/s up to 50 Mb/s, an OFDM system is viable solution for the following reasons:

- *Robustness against delay spread*: Data transmission in wireless environment experiences delay spreads of up to 800 ns which covers several symbols at baud rates of 10 Mb/s and higher. In a single carrier system an equalizer handles detrimental effects of delay spread. When delay spread is beyond 4 symbols, use of maximum likelihood sequence estimator structure is not practical due to its exponentially increasing complexity. Linear equalizers are not suitable for this application either since in a frequency selective channel it amounts to significant noise enhancement. Hence, other equalizer structures such as decision feedback equalizers are used. Number of taps of the equalizer should be enough to cancel the effect of inter-symbol interference and perform as a matched filter too. In addition, equalizer coefficients should be trained for every packet, as the channel characteristics are different for each packet. A large header is usually needed to guarantee the convergence of adaptive training techniques. A multi-carrier system is robust against delay spread and does not need a training sequence. Channel estimation is required however. A typical channel impulse response for indoor wireless LAN is shown in Figure 9.8.

- *Fall-back mode:* Depending on the delay spread for different applications a different number of carriers is required to null the effect of delay spread. The structure of FFT lends itself to simple fallback mode by using the butterfly structure of FFT as shown in Chapter 2.

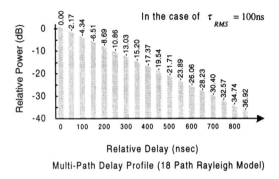

Figure 9.8. Channel impulse response for typical wireless LAN medium.

- *Computational efficiency:* Use of FFT structure at the receiver reduces the complexity to $N \log_2 N$. As the number of carriers grows the higher efficiency can be achieved. A typical OFDM Wireless LAN is shown in the Figure 9.9.

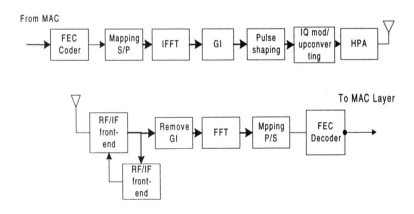

Figure 9.9. Wireless LAN system structure.

- *Fast synchronization:* OFDM receivers are less sensitive to timing jitter compared to spread spectrum techniques.

A typical OFDM wireless LAN transceiver is shown in Figure 9.9.

Transmitter Structure

WLAN system at 5 GHz supports variable data rates by using different modulation schemes and therefore requires variable coding gain for each data rate. Currently, using BPSK, QPSK, 16 QAM and 64 QAM modulation supports data rates of 6 Mb/s up to 54 Mb/s. Different coding rates for convolutional coding are proposed depending on required protection

Obviously, more crowded constellations need higher coding protection. A combination of error detection, such as CRC, and error correction, such as convolutional coding, in conjunction with interleaving are used to provide better coding gain in the presence of frequency selective fading. Convolutional coding is usually punctured to provide higher bandwidth efficiency. The number of carriers and modulation of data bits depends on the required data rates. Typical number of carriers is 48-52 and a guard band is added to control the spectrum shape. Additional guard interval in the time domain is used for pulse shaping and controlling the ISI. A typical frame format of OFDM WLAN is shown in Figure 9.10.

Preamble (AGC, Synchronization, Diversity)	Guard Interval	Training Seq. (Channel Estimation, Fine Sync)	Guard Interval	Service Data	Guard Interval	Data

Figure 9.10. Typical frame format for OFDM WLAN.

Pulse shaping has an important role in robustness of the system against phase noise and frequency offset as explained in Chapter 5. In addition, pulse shaping provides a smoother transmit signal spectrum which is critical in an OFDM system due to higher peak-to-average ratio. Transmitter output has to meet strict criteria of out-of-band leakage. A spectrum mask, such as the one shown in Figure 9.11, is usually defined to control the effect of bandwidth regrowth caused by amplifier clipping.

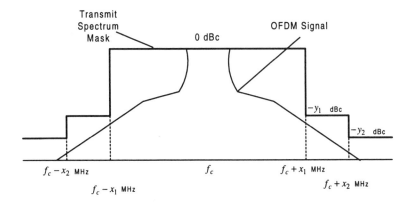

Figure 9.11. Transmitter spectrum mask.

A typical amplifier non-linearity model used in simulation and analysis is:

$$v_o = \frac{v_i}{(1+(\frac{|v_i|}{v_s})^{2p})^{1/2p}}$$

9.1

$$Back-off = -10\log\frac{Avg\{|v_i|^2}{|v_s|^2},$$

where p is smoothness factor and v_s is saturation voltage [7].

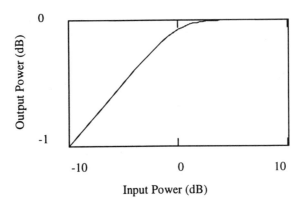

Figure 9.12: Nonlinear HPA model

Some typical degradation factors at the transmitter are briefly discussed here. Quadrature phase error and phase imbalance of transmitter modulation result in a C/N degradation and higher packet error rate. Another source of degradation is the fixed-point error caused by truncating the output of IFFT butterfly. After each stage of IFFT (or FFT) butterfly, the required number of bits grows. Truncation of bits can also increase the packet error rate. Maximum tolerable transmitter degradation is restricted by vector error requirements set forth in WLAN standards.

Currently, two frequency bands around 2.4 GHz and 5 GHz have been assigned to wireless bands. Target data rates of up to about 50 Mb/s and higher are also being considered for this system.

9.4 Receiver Structure

Acquisition

Initially, the receiver must detect the frame, adjust AGC to proper level, utilize any diversity capability, and adjust for coarse frequency offset. Fast synchronization is critical in this application. High phase noise and frequency offset of low cost receiver oscillators (with about ±10 ppm frequency stability) interfere with the synchronization process and require robust algorithm design.

A typical AGC block consists of a correlator followed by confirmation block. Since received signal level could vary significantly due to shadowing, a low threshold should be used followed by confirmation block, which utilizes the repetitive pattern of preamble sequence to reduce false alarm probability.

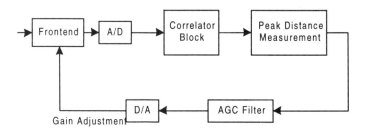

Figure 9.13. Frame synchronization and AGC.

Probability of false alarm and detection is a function of threshold Γ and signal to noise ratio. Probability of detection versus false alarm for different thresholds is referred to as Receiver Operation Characteristics (ROC) [4].

If more than one training sequence is used for synchronization the probability of false alarm decreases while probability of detection improves. The overall P'_F and P'_D will be

$$P'_F = \sum_{i=0}^{t} \binom{N}{i} p_f^i (1 - p_f)^{N-i}$$

$$P'_D = \sum_{i=t+1}^{N} \binom{N}{i} p_d^i P_m^{N-i},$$

9.2

where N is the number of correlations and t is the threshold determined by required probabilities of false alarm and detection.

If initial frequency offset is very high it may reduce the magnitude of peak and increase probability of false alarm. A frequency offset of Δf amounts to peak reduction of

$$\left| \sum_{n=1}^{N} e^{j2\pi n T \Delta f} \right| = \frac{\sin(\pi N T \Delta f)}{\sin(\pi T \Delta f)}.$$

9.3

In that case a peak detector can be implemented as a bank of correlators tuned to different frequency offset values. Therefore, frame synchronization, coarse frequency offset estimation and AGC can be implemented using a common structure of Figure 9.13.

Steady State

Once initial acquisition is accomplished, detection process begins. Detection process requires channel estimation, prefix removal, frequency domain equalization and soft decoding.

Channel estimation: Initial channel estimation block uses the preamble symbols. Typically, channel estimation block has the structure of Figure 9.14.

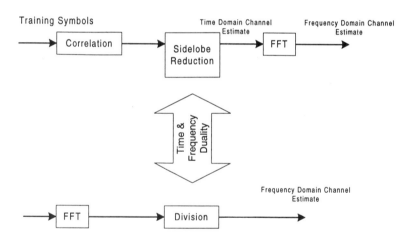

Figure 9.14. Channel estimation block.

After channel estimation and fine synchronization, guard intervals should be removed. The position of the OFDM block should be adjusted carefully to make sure that the effect of pulse shaping is preserved and interference from adjacent blocks are minimum.

Using a reliability index concept performs soft decoding of the demodulated signal. Due to the presence of an interleaver block a classic trellis decoding is not possible. We correspond a reliability parameter to each bit and use that in decoding. The reliability index for each bit is derived from the difference in metric value caused by bit flipping.

Typical degradation sources in the receiver are phase noise of the local oscillator, which was thoroughly discussed in previous chapters, and D/A quantization error. Overall, transmitter power amplifier non-linearity and receiver phase noise are dominant factors for total vector error.

References

1.   IEEE. "IEEE Standard 802.11, Part 11: Wireless LAN Medium Access Control (MAC) and Physical Layer (PHY) Specifications." Piscataway, NJ: IEEE, 1997.

2.   Draft Supplement to Standard for Telecommunication and Information Exchange Between Systems-LAN/MAN Specific Requirements-Part 11: Wireless Medium Access Control (MAC) and Physical Layer (PHY) Specifications. High Speed Physical Layer in the 5 GHz Band. Sept., 1998

3.   Stallings, W. *Data and Computer Communications*. New York: Maxwell MacMillan, 1991.

4.   Helstrom, C. W. *Statistical Theory of Signal Detection*. London: Pergamon, 1966.

7.   Bahai, A.R.S. "Estimation in Randomly Time-Varying Systems with Applications to Digital Communications." *Ph.D. Dissertation, Univ. of California at Berkeley*, 1993.

5.   Viterbi, A.J. *CDMA, Principles of Spread Spectrum Communication* Reading, MA: Addison Wesley, 1995.

6.   Takanashi, H. "Proposal of PHY Specification for 5 GHz Band." Proposal to IEEE 802.11 Committee, Jan 1998.

7.   Bogenfeld, E, et al. "Influence of Nonlinear HPA on Trellis-Coded OFDM for Terrestrial Broadcasting of Digital." *IEEE Proc. of Global Telecom Conf. (GLOBECOM '93)*; 1433-1438.

Chapter 10 *Digital Broadcasting*

10.1   Broadcasting of Digital Audio Signals

It has been recognized, particularly in Europe, that multi-carrier modulation is especially appropriate for broadcasting of digital signals. European standards bodies have adopted OFDM as the modulation choice for several variations of digital audio and digital television, for both terrestrial and satellite transmission [1-3]. We will concentrate here on one particularly interesting form of terrestrial digital audio broadcasting.

Substantial quality improvement can be obtained over conventional analog FM by using digital techniques, just as digital compact disk (CD) recording has widely supplanted analog disks and tapes. With modern compression techniques, high quality stereo audio can be produced at approximately 200 kb/s, an order of magnitude reduction from CD technology, which uses a straight PCM encoding without compression.

The European Digital Audio Broadcast (DAB) standard provides for three modes of operation. Mode 1 is for terrestrial Single Frequency Networks (SFN). Mode 2 applies to conventional terrestrial local broadcasting. Mode 3 is for satellite broadcasting. The ability to provide an

SFN is a unique capability of multi-carrier modulation, and the remainder of this section will discuss this application.

An SFN involves the provision of several geographically dispersed transmitters, all transmitting the same signal, synchronized in time. This permits coverage of a large area, and also fills in dead spots due to shadowing at a particular receiver location. Any receiver therefore receives a sum of signals from more than one transmitter, coincident in frequency, format, and time, except for differences in propagation delay. This is illustrated in Figure 10.1, showing multi-path propagation from several transmitters.

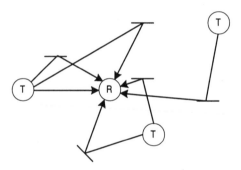

Figure 10.1. Received signal in an SFN.

Assuming that each path provides a Rayleigh distributed signal, the powers will add. If the spread of arrival times, which includes both multi-path spread and spread of propagation times from different transmitters, is less than the guard interval between OFDM symbols, then orthogonality among sub-carriers is preserved. If this is not the case, then degradation caused by interference among sub-carriers will result. This interference may be approximated as Gaussian noise, and is proportional to the degree that the inter-symbol interference duration exceeds the guard interval. As an example of a guard interval requirement, consider a receiver receiving direct signals from two transmitters 40 km apart. The time separation of the signals could be up to 133 μ sec. Considering the addition of multi-path from each transmitter, and reception from additional transmitters, the guard interval should be substantially greater than this value. Note that as in a cellular radio system, signals from far distant transmitters lead to system degradation.

A constraint in the other direction, due to Doppler shift and frequency variation in the receiver's local oscillator, sets a minimum spacing between sub-carriers, and thus a maximum symbol duration. For a vehicle travelling at 80 mph (31 m/s), the Doppler shift at 240 MHz can be up to 25 Hz. The sub-carrier spacing must be large compared to this value in order to minimize this source of inter-carrier interference. It is particularly important in an OFDM system that a good frequency lock be maintained in the receiver.

The Mode 1 standard meets these constraints by providing for 1536 sub-carriers with 1 kHz spacing. A guard interval, which employs a cyclic prefix, has a duration of 246 μ sec.

Since the useful symbol duration is 1 msec, the loss in spectral efficiency is approximately 20%. Each sub-carrier is modulated by π/4-DPSK, that is a 4-point constellation in which two bits of information are carried by differential phase modulation of

$$\pm\frac{\pi}{4} \text{ or } \pm\frac{3\pi}{4}. \qquad\qquad 10.1$$

A 2048-point IFFT is performed in the transmitter, treating the unused sub-carriers as zero, so the bandwidth remains 1.536 MHz. The signal is complex modulated, using I and Q components, onto an RF carrier in the range 175 - 240 MHz.

The transmitted OFDM signal carries several audio programs, each of which may be monophonic or stereo, and of varying quality. Each audio channel is encoded using sub-band encoding, at one of several allowed bit rates between 32 and 384 kb/s. That bit stream is then convolutionally encoded at rate of approximately ½ and interleaved. The bit streams are then time multiplexed and input to the OFDM modulator. The gross bit rate, as described below, is 2.3 Mb/s. Because of the variability of the original audio encoding, the number of channels carried is also variable. A typical system carries six programs each using 192 kb/s.

10.2   Signal Format

The construction of an overall transmitted DAB signal is shown in Figure 10.2.

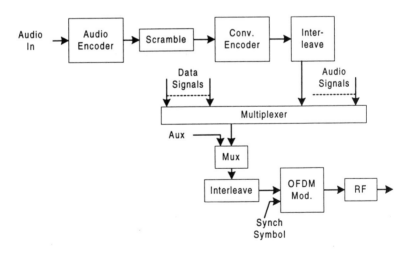

Figure 10.2. Block diagram of a DAB transmitter.

The audio encoder accepts digital samples of the audio input, which in general is 2-channel stereo. The samples occur at a 48 kHz. rate, 16 bits per sample, for a total rate of 768 kb/s. per channel. A sub-band encoding technique is used to reduce the rate to a lower value, depending on the desired subjective quality.

The output of the audio encoder is scrambled by modulo-2 addition with a pseudo-random sequence formed from a 9-bit feedback shift register. This is different from the more common scrambling technique of division by a randomizing polynomial. It does not suffer from error multiplication effects, but does require synchronization of the descrambling sequence at the receiver. The purpose of this scrambling is to insure that the final transmitted frequency spectrum is suitable dispersed. In addition, the standard allows for additional encryption scrambling if appropriate.

The scrambled data is then convolutionally encoded. This is essential for satisfactory operation over a radio channel. The first step is a mother code of rate ¼ with constraint length 7. That code is then subjected to puncturing to yield a rate of the form *8/n*, where *n* can be any integer value between 9 and 32. Each audio channel can have a different code rate, depending on the required error rate, to produce a given audio quality over the expected worst case channel. Typically a rate of approximately ½ is used. Interleaving by a factor of 16 is then applied.

Several audio channels are multiplexed and combined with other signals, which may include data channels in addition to audio. The signals are arranged in frames, as shown below. A frame consists of a time sequence of 2 synchronization symbols, 3 digital overhead symbols, followed by 72 symbols of multiplexed audio information. A symbol is a block of 3072 bits which will be mapped into an OFDM symbol. The length of a frame is therefore 77*1.246 ≈ 96 msec.

The overhead and information symbols are interleaved before assignment to sub-carriers. This provides further randomization of the signal. The synch symbols are added and the OFDM modulation performed.

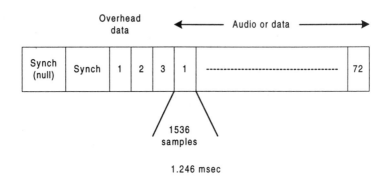

Figure 10.3. DAB frame structure.

The first synch symbol is null (no signal), so that course frame alignment can be obtained by simple envelope detection. The second synch symbol carries a fixed pattern. It is used both for fine acquisition and to provide phase reference for differential detection of the following symbols. The overhead bits constitute a fast information channel, which carries parameters necessary for decoding subsequent information.

The raw bit rate can be calculated as

$$\frac{72 \text{ symbols} * 1536 \text{ subcarriers} / \text{symbol} * 2 \text{ bits} / \text{subcarrier}}{.096 \text{ sec}} \qquad 10.2$$

$$= 2.3 \text{ Mb} / \text{s}.$$

As describer previously, each sub-carrier employs 4-point DPSK modulation. The differential modulation applies to the sequence of symbols of each sub-carrier, rather than among sub-carrier symbols as is sometimes proposed. Differential modulation eases the requirements of synchronization and simplifies receiver implementation, at the expense of approximately 2 dB in noise performance. For satisfactory performance, the OFDM symbol rate must be large compared to the maximum Doppler shift. An equivalent statement is that the OFDM symbol duration must be small compared to the time coherence of the channel.

10.3 Other Digital Broadcasting Systems

DAB in the U.S.A.

The system requirements for DAB in the U.S.A. are considerably different from those in Europe. Most importantly, it is desired to operate in the same 88 – 108 MHz band that is used for conventional analog FM broadcasting, using the same 200 kHz channels. This leads to a potentially severe problem of co-channel and adjacent channel interference between the digital and analog signals.

The problem of interference from the digital signals into analog receivers in reduced by transmitting the digital signals at a much lower power level than the analog signals. The FM capture effect then assists in rejection of the interference. Careful geographic frequency assignment and co-ordination is essential.

As for reception of the digital signal, interference from analog stations is a form of additive noise. Robust signal format and coding are essential. One suggested enhancement involves detecting an analog interferer and attempting to subtract it out [4, 5].

Standardization of the signal format is in progress, but some form of OFDM will most likely be adopted.

A standard similar to the DAB standard has been adopted in Europe for broadcasting of digital television signals. Using MPEG-2 coding, approximately 3 – 4 Mb/s are required for high quality SECAM or PAL programming, while up to 20 Mb/s are needed for future high definition television.

A bandwidth of 8 MHz is used, as in PAL or SECAM, to carry either several standard channels or one high definition channel. The same SFN scheme as in DAB will be deployed. In this case 8000 sub-carriers spaced approximately 1 kHz apart are used. Larger constellation sizes, up to 64 points, provide the higher bit rates.

The large constellation sizes require more robust techniques than are used in DAB [6, 7]. Modulation is coherent as opposed to differential. A concatenated inner convolution code and outer block code are employed. Several pilot tones are transmitted for more precise equalization and synchronization.

10.4 Digital Video Broadcasting

Digital video broadcasting services cover more than digital television. The wireless video concept includes applications such as multimedia mobile and wireless data services. In addition, DVB systems can be integrated with other wireless services such as cellular systems to provide a highly asymmetric data access system which is desirable for new wireless Internet access.

Source coding is one of the most critical components of DVB systems. Different data rates are required for the variety of services offered by DVB. Quality television requires about 216 Mb/s and stereo audio requires about 1536 kb/s with no source coding applied. MPEG 2 utilizes a compression algorithm to remove spatial, temporal and statistical redundancy of images, and sub-band coding techniques in conjunction with psycho-acoustic models to reduce the bit rate. The total bit rate, video and audio is reduced to approximately 3.5 Mb/s.

The choice of Coded Orthogonal Frequency Division Multiplexing (COFDM) for DVB-T applications is based on several design factors:

- *Interference rejection:* Co-existence of digital broadcasting systems with current analoge services requires a system design which restricts interference into analoge systems and is robust against narrow band interference. The former can be met by reducing power level and utilizing coding techniques of digital communications to achieve required bit error rates. The latter can be met by multi-carrier modulation efficiently due to its robustness against narrow band interference such as VSB-AM.

- *Multi-path effect:* Multi-path propagation is a major impairment of a terrestrial broadcasting channel. The SFN architecture of DVB produces strong echo (0 dB) generated by adjacent transmitters. Presence of strong echo-like interference requires a robust modulation scheme. COFDM is again suitable to combat against strong multi-path as long as the guard interval length is properly designed. Larger guard intervals, as explained before, allows larger distance between transmitters at the expense of reduced bandwidth utilization. Proper positioning of pilot signals in time and frequency slots provides an acceptable performance in higher Doppler conditions such as mobile multimedia terminals.

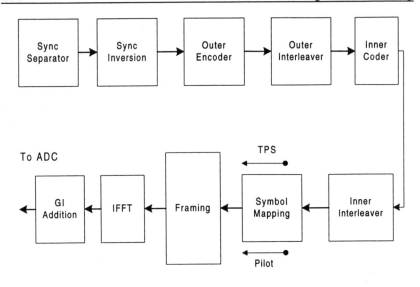

Figure 10.4. DVB transmitter structure.

After down conversion at the receiver's analog front-end and sampling, a digital AGC is used to adjust the VGA gain in the front-end circuit. A coarse frequency offset correction is necessary to reduce the inter-carrier interference. Frequency offset estimation uses the fixed pilots for initial estimation and scattered pilots for further fine synchronization. Time synchronization is initially performed by using a correlator structure and further refinement is done by methods explained in Chapter 5 for fine synchronization using the output of the OFDM demodulator. Channel estimation is required for detection and soft decoding reliability measure. Pilot symbols are used for channel estimation, and by using time and frequency interpolation we can obtain a channel estimate for all sub-carriers adaptively. Performance of the inner decoder is significantly enhanced by using a soft decoder and a reliability measure derived from channel estimation. A typical receiver is shown in Figure 10.5.

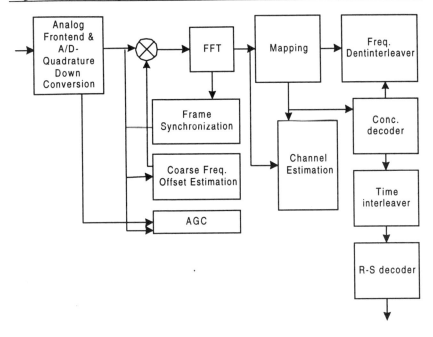

Figure 10.5. DVB receiver architecture.

References

1. Alard, M., Lassalle R. "Principles of Modulation and Channel Coding for Digital Broadcasting for Mobile Receivers." *EBU Review*; Aug 1987; 224: 168–190.

2. Taura, K., Tsujishita M., Takeda M., Kato H., Ishida M., Ishida Y. "A Digital Audio Broadcasting (DAB) Receiver." *IEEE Trans. Consumer Elect.*; Aug 1996; 42; 3: 322-327.

3. Kuchen, F.; Becker T.C., Wiesbeck W. "Analytic Bit Error Rate Determination for Digital Audio Broadcasting." *IEE Int. Broadcasting Conv.*; Sep 1996: 261-266.

4. Leclerc, M, Scalart P., Fautier P., Huynh H.T. "Performance Analysis of an In-Band COFDM/FM Digital Audio Broadcasting System." *IEEE Trans. Broadcasting*; Jun 1997; 43; 2: 191-198.

5. Kroeger, B.W., Peyla P.J. "Compatibility of FM Hybrid In-Band On-Channel (IBOC) System for Digital Audio Broadcasting." *IEEE Trans. Broadcasting*; Dec 1997; 43; 4: 421-430.

6. Mignone, V., Morello A., Visintin M. "An Advanced Algorithm for Improving DVB-T Coverage in SFN." *IEEE Int. Broadcasting Conv.*; Sep 1997: 534-540.

7. Sari, H., Karam G., Jeanclaude I. "Transmission Techniques for Digital Terrestrial TV Broadcasting." *IEEE Commun. Mag.*; Feb 1995; 33; 2: 100-109.

Chapter 11 *Future Trends*

Multi-carrier communications are currently in a state of very rapid development, both in the introduction of new techniques and in application to a wide variety of channels and services. Advances in algorithmic development and in the capabilities of signal processing devices will surely lead to substantial improvement in performance of multi-carrier systems. This in turn will lead to further application in systems where the unique advantages of multi-carrier modulation can be applied, while overcoming its drawbacks, at reasonable cost.

11.1 Comparison with Single Carrier Modulation

In Reference [1] a comparison was made of the relative advantages of OFDM and single carrier modulation in the ADSL application. The table below indicates which system is favored for each of several issues.

Issue	Single Carrier	Multi-carrier	Equivalent
Performance in Gaussian noise			X
Sensitivity to impulse noise (uncoded)		X	
Sensitivity to narrowband noise (uncoded)	X		
Sensitivity to clipping	X		
Sensitivity to timing jitter and phase noise	X		
Latency (delay)	X		
Computations per unit time		X	
Cost and power consumption in analog sections	X		
Adaptability of bit rate		X	

Table 1: Relative advantages of single carrier and multi-carrier modulation.
An "X" denotes the system with better performance or lower cost.

The comparisons also apply for environments other than ADSL. As shown in Chapter 3, bandwidth efficiency and performance in the presence of colored Gaussian noise ideally are equal, if the single carrier system uses optimum decision feedback equalization and the multi-carrier system employs optimum bit loading. The relative computation per unit time is less for multi-carrier in the ADSL environment where the long equalizer needed to correct the highly dispersive channel response in a single carrier implementation requires more computation than the short equalizer and FFT in multi-carrier. The higher cost and power drain of analog components for multi-carrier, particularly the transmitter power amplifier, results from the necessity to keep clipping at a minimum. The sensitivity to narrowband noise and jitter are due to the overlapping and slowly decaying sub-carrier spectra in multi-carrier modulation.

Future enhancements to OFDM give promise of reversing some of the cases in which OFDM compares unfavorably, in particular those associated with clipping and the overlapping sub-carrier spectra in current OFDM implementations.

11.2 Mitigation of Clipping Effects

In Chapter 4 it was shown that the error probability resulting from clipping may be quite tolerable in most cases. However the need to limit the cost, and even more importantly the power drain, of components such as the transmitter power amplifier provides an incentive to keep the peak-to-average power ratio as small as possible. Several approaches have been proposed to accomplish this.

Probably the simplest solution is to allow clipping at a relatively low level, and use standard error control coding to correct the resultant errors. Of course this also reduces errors caused by other degradations, in particular additive noise. In [2] it is shown that this simple approach may compare favorably with other techniques when the clipping level is not reduced to too low a value. Another receiving scheme involves attempting to reconstruct the clipped signal [3].

A large number of proposals involve altering the transmitted signal so as to keep the peak value low, and in most cases notifying the receiver of the alteration. Because the probability of a clip is relatively low in any case, the side information required to perform this notification can be quite low compared with the actual data [4]. One relatively simple approach is to buffer each OFDM symbol before transmission and to detect if the clipping level would be exceeded. If so, the level is reduced and the reported to the receiver, possibly on a sub-carrier dedicated to this purpose.

Other techniques [5, 6, 7] use block codes to produce more than one mapping of input data to OFDM symbols, and transmit the one with lowest peak. The mappings typically involve different rotations of the sub-carrier constellations. Side information is sent to the receiver as before to report which mapping was used. In practice only one or two bits of side information per OFDM symbol are needed. If unused sub-carriers are available, then they could be modulated in a way that reduces the peak value of each OFDM symbol. In this case no side information need be transmitted because those sub-carriers are ignored in any case.

In Reference [6], it is shown that that for an error probability of 10^{-7}, the peak-to-average ratio can be reduced from 13.3 dB to 11.1 dB by the use of two mappings, or down to 9.5 dB with four mappings.

Reduction of the peak value of an OFDM signal is now an active field of research, involving trade-offs among performance, spectral efficiency, and complexity of implementation. Incorporation of some of these techniques in certain applications of OFDM may be expected in the near future.

11.3 Overlapped Transforms

When the time window used in modulation is a square one equal to the OFDM symbol duration, the spectrum of each sub-carrier is of the *sinc* form. Each sub-carrier spectrum therefore overlaps many of the others. This leads to high inter-channel interference when orthogonality is lost for any reason, such as inter-symbol interference, frequency offset, or phase noise. In addition, the susceptibility to narrow-band interference can be severe in that many sub-channels could be affected by a single interferer. The problem can partially be relieved by the use of windowing as described in Chapter 2. However there is only the duration of the cyclic prefix to provide room for the window shaping, so that substantial overlap of sub-carrier spectra will remain.

Much better spectral shaping can be achieved if processing is no longer constrained to the duration of a block corresponding to a single OFDM symbol. If several blocks are allowed to overlap in time, then a very gradual windowing function can be introduced. Reference [8] describes the performance that can be achieved by a technique called Discrete Wavelet Multi-Tone (DWMT) modulation. In particular, for an overlap of 8 blocks, the first sidelobe of each sub-carrier spectrum is reduced from -13.5 dB in DMT to -45 dB. Performance in noise is substantially unchanged, and sensitivity to various inaccuracies is greatly reduced. Furthermore, no cyclic prefix is needed, leading to maximum spectral efficiency. However DWMT requires substantially greater computational complexity, and latency is increased by the degree of overlap. The DWMT transmitter is shown in Figure 11.1.

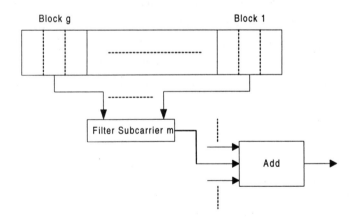

Figure 11.1. DWMT transmitter.

As in DMT, the filter for sub-carrier n, $F_n(z)$ is centered at frequency n/T with nulls at the center frequencies of other sub-carriers. Unlike DMT, its impulse response is of length gT, where $g > 1$. The set of filters performs a wavelet transform that is not described in the reference. The equivalent of the filter bank can be implemented by a fast algorithm similar to the FFT. The impulse responses are orthogonal both over frequency index and over time shifts.

$$\sum_k f_k^n \, f_{k-iN}^m = K\delta_i \delta_{n-m}.$$ 11.1

The filters are also chosen so that their spectra decay very rapidly. Overlap of sub-carriers is therefore greatly reduced. Increasing g enhances this spectral confinement.

With input symbols D_m the line signal is of the form

$$d(z) = \sum_{n=0}^{N-1} D_n(z) F^m(z).$$ 11.2

If there is any dispersion at all over the channel, then unlike in DMT with cyclic prefix, orthogonality among the sub-carriers is lost. This is corrected by means of a matrix equalizer whose size can be kept relatively small due to the very low spectral overlap of sub-carriers spaced a moderate number apart. The matrix equalizer must not only correct for inter-symbol interference in each sub-carrier, but also inter-channel interference among the sub-carriers. Because the frequency overlap among the sub-carriers decays very quickly, only immediately adjacent sub-carriers typically need be accounted for.

The DWMT receiver is illustrated below. The incoming signal may first be partially equalized by an optional pre-equalizer. This can relieve the computational burden of the matrix post-equalizer. A wavelet transform, matched to the transmitter transform, is performed by the equivalent of the filter bank shown. Again a fast algorithm can be used. This is followed by the matrix equalizer which includes adjacent sub-carrier taps as well as the usual equalization for each sub-carrier.

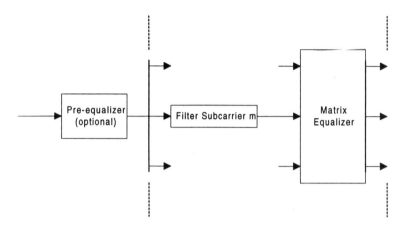

Figure 11.2. DWMT receiver.

The DWMT technique clearly requires more computation than DMT. Reference [8] reports that the complexity in a modulator and demodulator is increased by a factor of approximately $1.25 \dfrac{1 + g + \log_2 N}{\log_2 N}$, which is not especially large.

The DWMT technique is a promising one. Not only does it reduce susceptibility to narrowband interference and loss of orthogonality, but the increase in spectral efficiency due to elimination of guard times between OFDM symbols may well be worth the additional cost in many applications. However the increase in latency must be considered in some applications.

11.4 Combined CDMA and OFDM

Several proposals have been made to combine the use of CDMA for wireless multiple access with the desirable properties of OFDM [9, 10]. The longer symbol interval in OFDM greatly eases the problem of acquiring synchronization of the spreading code in the receiver. The sub-carriers are spaced sufficiently close such that each sub-carrier constitutes a frequency non-selective fading channel. This eliminates the need for RAKE combining.

Figure 11.3 shows one scheme. The same data symbol, multiplied by a different bit of the spreading code $\{c\}$, modulates each sub-carrier. This in effect provides spreading in the frequency domain. The transmitted signal element duration is the same as the original symbol duration, rather than the chip duration as in ordinary CDMA. Synchronization is therefore a much simpler process. Provided that adequate guard intervals, using cyclic prefixes, are used, multi-path correction is easier and better than RAKE combining in a single carrier system. This OFDM-CDMA technique provides frequency diversity. Best performance in this environment is achieved through the use of optimum combining in the choice of weighting factors $\{q\}$ that are applied in the receiver after demodulation and despreading.

In the downstream direction (from a central station to the mobile) multiple access is achieved by superimposing signals for different users on the same sub-carriers using a set of orthogonal spreading codes such as Walsh functions. In the upstream direction a set of spreading codes, such as Gold codes, with good auto-correlation and cross-correlation properties, are employed. However in this case a multi-user detection scheme in the receiver is essential because the asynchronous arrival times destroy orthogonality among the OFDM sub-carriers.

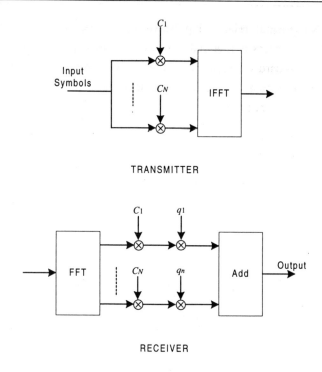

Figure 11.3. OFDM-CDMA system with frequency spreading.

The above description assumes that the number of spreading chips per symbol, or the processing gain, is equal to the number of sub-carriers. This is not necessary. If more sub-carriers are desirable, either to achieve flat spectra or to achieve greater spectral efficiency with a given guard interval, then more than one input symbol may be spread to form each OFDM symbol.

Another form of combined CDMA-OFDM involves feeding different spread input symbols to the sub-carriers, as illustrated in Figure 11.4. The spreading codes for a given user may all be the same as shown, or could be different. No integral relationship between the processing gain and the number of sub-carriers is required. This process uses spreading in the time domain as in standard CDMA, but at a rate much lower by a factor of *1/N*. For the same reasons as in the first scheme, code acquisition is eased and RAKE combining is not needed.

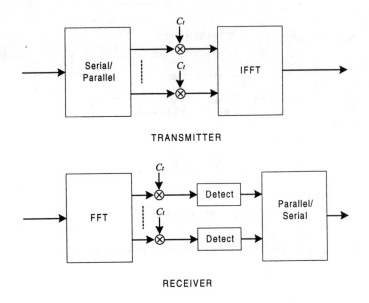

Figure 11.4. OFDM-CDMA system with time spreading.

Other variations have been proposed, including very high rate modulation so that sub-channel spectra severely overlap. RAKE type techniques are required to receive this signal.

A different interpretation of OFDM-CDMA using a duality principle in digital communications is presented in [10]. Using the duality principle, we can design the receiver as dual of a single carrier CDMA rake receiver. This duality is shown in the following figure:

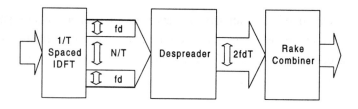

Figure 11.5. Discrete time MC-SS rake in frequency domain.

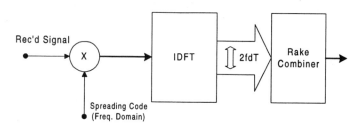

Figure 11.6. Simplified rake receiver.

11.5 Advances in Implementation

"Moore's Law" concerning the exponential increase in the number of elements that can be included on a chip is expected to hold for the next few years. One consequence will be the availability of general purpose signal processing chips that can perform advanced algorithms for fairly high rate OFDM systems. Another is the availability of circuitry that is dedicated to performing common functions efficiently, either as part of a signal processing chip or as an auxiliary chip. Functions now available include FFTs, Viterbi decoders, Reed-Solomon coders and decoders, etc.

Implementation of the butterfly operation of FFT in a modern DSP is significantly more efficient than before. Execution of a 1024-point bit reversed complex Fast Fourier Transform in a new DSP takes less than 50 microseconds.

Analog/Digital converters have been critical components in terms of cost in the past. Because of the high peak-to-average ratio of OFDM signals, this has been a serious issue for such systems. However recent advances in signal conversion devices have reduced this concern. Availability of high speed A/D devices with high spurious-free dynamic range (SFDR) and signal-to-noise ratio (SNR) at reasonable cost has resulted in further research and development in *Digital Radio* architecture. By using a combination of high speed A/D and digital down conversion, complexity of an analog front-end is reduced and the system architecture is more flexible. Programmability of digital radio is an attractive feature for multi-mode and multi-band wireless application.

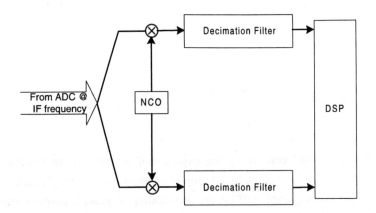

Figure 11.7. IF sampling and digital radio down conversion.

Characteristics of power amplifiers are important for OFDM system physical layer performance such as bit error rate level, as discussed in Chapter 4, and for power consumption requirements. Power amplifiers are the largest source of battery drain in wireless systems. Recent advances in GaAs MMIC has resulted in highly efficient wideband power amplifiers suitable for OFDM applications.

References

1. Saltzberg, B.R. "Comparison of Single Carrier and Multitone Digital Modulation for ADSL Applications." *IEEE Commun. Magazine*; Nov 1988; 36; 11: 114-121.

2. Wulich, D., Goldfeld L. "Reduction of Peak Factor in Orthogonal Multicarrier Modulation by Amplitude Limiting and Coding." *IEEE Trans. Commun.*; Jan 1999; 47: 18-21.

3. Kim, D., Stuber S.L. "Clipping Noise Mitigation for OFDM by Decision-Aided Reconstruction." *IEEE Commun. Let.*; Jan 1999; 3; 1: 4–6.

4. Starr, T., Cioffi J.M., Silverman P. *Understanding Digital Subscriber Line Technology.* Upper Saddle River, NJ: Prentice Hall, 1999, 236-242.

5. Mestdagh, D.J.G., Spruyt P.M.P. "A Method to Reduce the Probability of Clipping in DMT-Based Transceivers." *IEEE Trans. Commun.*; Oct 1996; 44: 1234-1238.

6. Baum, R. W., Fischer R. F. H., Huber J. B. "Reducing the Peak-to-Average Ratio of Multicarrier Modulation by Selective Mapping." *Electronic Letters*; Oct 1996; 32: 2056-2057.

7. Shepard, S., Orriss J., Barton S. "Asymptotic Limits in Peak Envelope Reduction by Redundancy Coding in OFDM Modulation." *IEEE Trans. Commun.*; Jan 1998; 46: 5-10.

8. Sandberg, S.D., Tzannes M.A. "Overlapped Discrete Multitone Modulation for High Speed Copper Wire Communication." *IEEE J. on Sel. Areas in Commun.*; Dec 1995; SAC-13; 9: 1571-1585.

9. Hara, S., Prasad R. "Overview of Multicarrier CDMA." *IEEE Commun. Magazine*; Dec 1997; 35; 12: 126-133.

10. Bahai, A., Fettwies G. " Results on MC-CDMA Receiver Design." *IEEE Int'l Conf. Comm*; Jun 1995: 915-918.

Index

A

Adaptive · 3, 110, 116, 133, 154, 171, 181, 182
ADSL (Asymmetric Digital Subscriber Line) · vi, 14, 55, 68, 72, 80, 115, 124, 125, 126, 159, 161, 170, 172, 203, 204, 215
Aliasing · 23, 28, 76
Alphabet · 2, 3, 4, 139
ARMA model · 109, 112, 113, 119, 127, 130, 131, 132, 133
Auto-correlation · 42, 107, 210

B

Bandwidth regrowth · 76, 185
Bit Allocation · 22, 46, 171
Bridged tap · 163
Bussgang's theorem · 64, 81

C

CDMA (Code Division Multiple Access) · 190, 210, 212, 215
Channel estimation · 29, 180, 182, 189, 200
Clipping · vi, 57, 58, 60, 61, 62, 63, 64, 66, 69, 71, 72, 73, 74, 75, 76, 77, 78, 79, 80, 185, 204, 205, 215
Coding, block · 136, 138, 140
Coding, concatenated · 136, 147, 148, 153, 198
Coding, convolutional · 136, 142, 143, 144, 146, 147, 148, 149, 151, 153, 157, 184
Coding, Reed-Solomon · 139, 148, 156, 171, 172, 213
Coding, Trellis · 47, 136, 148, 149, 150, 151, 152, 157, 171, 172, 190
Coding, turbo · 136, 153, 154, 155, 156, 157
Coherence · 30, 34, 99, 139, 197, 198

Conjugate appending · 13, 24, 137, 140, 171
CRC (Cyclic redundancy check) · 140, 141, 171, 172, 184
Crosstalk · 47, 135, 164, 165, 166, 167, 168, 169, 170, 173
Cyclic extension · 12, 21, 27, 28, 29, 87, 116, 117, 118, 123, 124, 171, 193, 207, 208, 210

D

DAB (Digital Audio Broadcast) · 191, 194, 196, 197, 198, 201
Decimation in frequency · 26
Decimation in time · 21, 26
DFE (Decision Feedback Equalization) · 51, 52, 55, 104, 106, 109, 110
DMT (Discrete Multi-Tone) · 12, 13, 14, 45, 80, 109, 110, 119, 160, 207, 208, 209
Doppler · 34, 193, 197, 199

E

Echo · 104, 120, 121, 122, 123, 124, 125, 126, 133, 163, 171, 199
Echo cancellation · 104, 120, 121, 123, 124, 125, 126, 133, 171
Equalization · vi, 8, 14, 31, 55, 103, 104, 109, 115, 116, 117, 118, 119, 120, 121, 122, 123, 126, 129, 133, 181, 189, 198, 204, 209
Error probability · 4, 42, 43, 44, 46, 47, 48, 50, 51, 53, 54, 58, 66, 68, 69, 70, 71, 72, 73, 74, 75, 79, 118, 135, 137, 147, 149, 152, 171, 205, 206

F

FDM (Frequency Division Multiplexing) · 7
FFT (Fast Fourier Transform) · 11, 12, 13, 14, 17, 18, 19, 21, 23, 25, 26, 27, 29, 84, 89, 91, 104, 119, 182, 183, 186, 204, 208, 214
FFT butterfly · 26, 182, 186, 214
FFT, partial · 25, 26, 27, 29
Frame synchronization · 89, 90, 187, 188
Frequency offset · 29, 83, 85, 86, 87, 88, 89, 90, 91, 92, 93, 97, 101, 180, 185, 187, 188, 200, 206

G

Gap · 51
Gating · 9
Gramm-Schmidt · 64
Gray coding · 44, 72
Guard interval · 79, 87, 103, 109, 184, 189, 192, 193, 199, 210, 211

I

IEEE 802.11 Standard · 178, 181, 190
Interference · 19, 29, 58, 76, 79, 83, 85, 86, 98, 99, 100, 101, 105, 109, 118, 135, 160, 167, 169, 176, 181, 189, 192, 193, 197, 199, 200, 206, 209
Interference, inter-channel · 85, 86, 87, 89, 103, 206, 208
Interference, inter-symbol · 3, 4, 8, 9, 27, 28, 29, 83, 85, 86, 87, 103, 104, 105, 106, 108, 109, 123, 182, 184, 192, 206, 208

Interleaving · 14, 136, 140, 143, 146, 148, 153, 154, 155, 157, 171, 184, 194, 195, 196
ISO layers · 175, 176

J

Jitter · 83, 85, 101, 184, 204, 205

K

Kineplex · 8, 9

L

LAN (Local Area Network) · 175, 176, 177, 181
LAN, Wireless · vi, 79, 80, 89, 175, 176, 177, 178, 180, 181, 182, 183, 184, 186, 190

M

Matched filter · 4, 31, 104, 105, 108, 182

N

Noise · 2, 4, 5, 6, 8, 19, 22, 30, 31, 38, 41, 42, 43, 44, 45, 46, 47, 48, 51, 52, 54, 58, 63, 65, 72, 73, 74, 75, 76, 81, 92, 93, 94, 96, 97, 99, 100, 104, 105, 106, 107, 118, 119, 121, 131, 135, 138, 139, 146, 149, 163, 164, 169, 182, 190, 192, 197, 204, 205, 207, 214, 215
Nyquist criterion · 3, 9, 11, 16

O

OFDM · v, vi, 5, 7, 8, 10, 11, 12, 13, 14, 15, 17, 18, 20, 21, 25, 28, 29, 32, 34, 35, 37, 42, 45, 46, 47, 55, 57, 58, 64, 66, 70, 72, 74, 76, 79, 80, 83, 84, 85, 86, 87, 88, 89, 97, 98, 100, 101, 103, 109, 116, 117, 118, 119, 122, 123, 124, 125, 126, 135, 136, 137, 146, 148, 152, 153, 155, 160, 170, 181, 182, 183, 184, 185, 189, 191, 192, 193, 195, 196, 197, 198, 200, 203, 205, 206, 207, 209, 210, 211, 213, 214, 215

P

PAM (Pulse Amplitude Modulation) · 2, 3, 5, 6, 42
Peak-to-average ratio · 59, 76, 185, 205, 206, 214, 215
Phase noise · 83, 93, 94, 96, 97, 98, 99, 100, 101, 185, 187, 190, 204, 206
Phase-locked loop · 94, 95, 96, 97, 101
Pilot · 29, 30, 89, 91, 98, 198, 199, 200
Power allocation · 53
Power amplifiers · 57, 59, 60, 76, 79, 190, 204, 205, 215

Q

QAM (Quadrature Amplitude Modulation) · 5, 6, 7, 9, 10, 12, 16, 42, 43, 44, 45, 57, 150, 152, 184
QAM, staggered · 11

R

Rayleigh fading · 20, 35, 61, 62, 63, 68, 69, 73, 74, 78, 146, 147, 192
Roll-off · 4, 11, 85, 87

S

Scattering function · 33
SFN (Single Frequency Network) · 14, 191, 192, 198, 199, 202
Shaping gain · 45
Synchronization · vi, 14, 23, 83, 88, 89, 91, 154, 171, 180, 181, 184, 187, 188, 189, 192, 195, 196, 198, 200, 210

T

Time-varying channels · 19, 29, 32, 33, 34, 35, 36, 37, 39, 190

V

Viterbi algorithm · 143, 144, 146, 147, 148, 151, 152, 154, 155, 190, 213

W

Water pouring · 48, 49
Windowing · 18, 19, 28, 29, 45, 84, 86, 87, 110, 113, 179, 206, 207
Wireless · v, vi, 14, 15, 21, 22, 32, 59, 72, 76, 79, 80, 83, 89, 93, 135, 155, 175, 176, 178, 179, 180, 181, 182, 183, 184, 186, 190, 198, 210, 214, 215
Wire-Pair channel · vi, 47, 120, 160, 161, 162, 163, 169, 170